Bastian Herzog

Benzotriazole and Sulfamethoxazole

Bastian Herzog

Benzotriazole and Sulfamethoxazole

Biodegradation of polar, non-adsorptive xenobiotic micropollutants with activated sludge communities and pure cultures

Südwestdeutscher Verlag für Hochschulschriften

Impressum / Imprint
Bibliografische Information der Deutschen Nationalbibliothek: Die Deutsche Nationalbibliothek verzeichnet diese Publikation in der Deutschen Nationalbibliografie; detaillierte bibliografische Daten sind im Internet über http://dnb.d-nb.de abrufbar.
Alle in diesem Buch genannten Marken und Produktnamen unterliegen warenzeichen-, marken- oder patentrechtlichem Schutz bzw. sind Warenzeichen oder eingetragene Warenzeichen der jeweiligen Inhaber. Die Wiedergabe von Marken, Produktnamen, Gebrauchsnamen, Handelsnamen, Warenbezeichnungen u.s.w. in diesem Werk berechtigt auch ohne besondere Kennzeichnung nicht zu der Annahme, dass solche Namen im Sinne der Warenzeichen- und Markenschutzgesetzgebung als frei zu betrachten wären und daher von jedermann benutzt werden dürften.

Bibliographic information published by the Deutsche Nationalbibliothek: The Deutsche Nationalbibliothek lists this publication in the Deutsche Nationalbibliografie; detailed bibliographic data are available in the Internet at http://dnb.d-nb.de.
Any brand names and product names mentioned in this book are subject to trademark, brand or patent protection and are trademarks or registered trademarks of their respective holders. The use of brand names, product names, common names, trade names, product descriptions etc. even without a particular marking in this works is in no way to be construed to mean that such names may be regarded as unrestricted in respect of trademark and brand protection legislation and could thus be used by anyone.

Coverbild / Cover image: www.ingimage.com

Verlag / Publisher:
Südwestdeutscher Verlag für Hochschulschriften
ist ein Imprint der / is a trademark of
OmniScriptum GmbH & Co. KG
Heinrich-Böcking-Str. 6-8, 66121 Saarbrücken, Deutschland / Germany
Email: info@svh-verlag.de

Herstellung: siehe letzte Seite /
Printed at: see last page
ISBN: 978-3-8381-3826-8

Zugl. / Approved by: München, TU, Diss., 2013

Copyright © 2014 OmniScriptum GmbH & Co. KG
Alle Rechte vorbehalten. / All rights reserved. Saarbrücken 2014

*What we call the beginning is often the end.
And to make an end is to make a beginning.
The end is where we start from.*

T. S. Eliot

ABSTRACT

The polar, xenobiotic micropollutants 1-H-benzotriazole, 4- and 5-tolyltriazole (BTri, 4-TTri, 5-TTri, summarized as BTs), and the sulfonamide antibiotic sulfamethoxazole (SMX) are due to their widespread applications almost omnipresent in aquatic systems, including drinking water. Therefore, this work was conducted to evaluate the biodegradation potential of activated sludge communities (ASCs) and pure cultures towards these pollutants.

To screen a high number of different setups for BTs and SMX biodegradation within a short time, an easy to perform and very cost-efficient UV-absorbance-based measurement (UV-AM) system was established allowing monitoring biodegradation under laboratory conditions and at xenobiotic concentrations above 1.0 mg L^{-1}.

In a first attempt, three different wastewater treatment plants were monitored for their BTs concentrations and showed 5-TTri to be best removed with a mean of 75% followed by BTri with up to 45% and worst for 4-TTri with up to 15% only. Analysis of the removal in different treatment stages showed that 5-TTri was mainly biodegraded in the aeration tanks while BTri and 4-TTri were equally removed across all treatment stages. Measurements of the receiving rivers up- and down-stream the WWTPs proved the latter to be a point source for benzotriazoles in the aquatic environment.

Up to 30 mg L^{-1} BTri and 5-TTri were biodegraded in non-acclimated ASCs under aerobic conditions at 20°C in up to 49 and 7 days, respectively, but not under denitrifying, sulfate reducing or anaerobic conditions. 4-TTri was biologically stable under all applied conditions. Acclimation significantly improved BTri and 5-TTri biodegradation. While BTri, after acclimation, was removed within 7 days, 5-TTri removal could be reduced to 4 days. Nutrient supply crucially improved biodegradation, especially nitrogen concentrations were observed to play a critical role for biodegradation. While carbon showed no such effect, additionally supplied nitrogen increased biodegradation that became faster and more efficient.

Further experiments regarding the enhancement of 5-TTri biodegradation by optimizing nutrient supply and acclimation were performed. Acclimation and nutrient supply were identified as main factors to improve the biological removal of 5-TTri. Acclimation over several generations optimally adjusted the ASC to utilizing 5-TTri immediately after inoculation and with high biodegradation rates up to 5.2 mg L^{-1} d^{-1}. Additional experiments, performed with an extract from autoclaved activated sludge supernatant to simulate wastewater nutrient conditions and with specific nitrogen compounds, significantly increased 5-TTri biodegradation rates up to 5.0 mg L^{-1} d^{-1} without any acclimation. These experiments indicated that 5-TTri removal is strongly dependent on nitrogen supply and thus might start with benzene ring cleavage, necessitating nitrogen supply. Acclimated ASCs that showed best biodegradation potential were characterized by DGGE (denaturing gradient gel electrophoresis), metagenomic and metatranscriptomic analysis. DGGE revealed a low biodiversity in the ASC consisting of four dominant species: *Aminobacter* sp.,

Flavobacterium sp., *Hydrogenophaga* sp. and *Pseudomonas* sp.. In contrast, metagenomic analysis showed a higher diversity in the ASCs with the most prominent species being *Mesorhizobium* spp., *Pseudomonas* spp., *Acidovorax* spp. and *Hydrogenophaga* spp.. Metatranscriptomic analysis to reveal the ASCs' metabolic activity revealed *Pseudomonas* spp., *Hydrogenophaga* spp. and *Acidovorax* spp. as the most likely species contributing to 5-TTri biodegradation as they were found with high activities in the biodegrading setups.

A different approach aimed at the evaluation of SMX biodegradation. Nine different bacteria species, capable of SMX biodegradation, were isolated from SMX-acclimated ASCs. 16S rRNA gene sequencing revealed five *Pseudomonas* spp., one *Brevundimonas* sp., one *Variovorax* sp. and two *Microbacterium* spp.. These cultures, incubated in media containing 10 mg L^{-1} SMX and different carbon and nitrogen concentrations, revealed biodegradation rates up to 2.5 mg L^{-1} d^{-1} in complex media under aerobic conditions and room temperature confirming that readily degradable energy sources were crucial for efficient SMX biodegradation. Moreover, media without any carbon and nitrogen supplementation proved the organisms' ability to utilize SMX as sole energy and nutrient source.

KURZFASSUNG

Die polaren Xenobiotika 1-H-Benzotriazol (BTri) sowie 4- und 5-Tolyltriazol (4-TTri, 5-TTri, zusammen als BTs bezeichnet) und Sulfamethoxazol (SMX, Sulfonamid-Antibiotikum) werden vielfältig eingesetzt und finden sich in nahezu allen Oberflächengewässern und im Trinkwasser. Die vorliegende Arbeit behandelt das biologische Abbaupotential von Belebtschlammbiozönosen als auch von Reinkulturen gegenüber den erwähnten Stoffen.

Um eine große Anzahl verschiedener Abbauversuche überwachen und kontrollieren zu können, wurde eine leicht anwendbare und kosteneffiziente UV-Absorptions-basierte Messmethode (UV-AM) etabliert. Sie ermöglicht die Darstellung des biologischen Abbaus der getesteten Xenobiotika unter Laborbedingungen und Konzentrationen größer 1,0 mg L^{-1}. Dadurch wurde es möglich weiterführende Experimente mit geringem Kostenaufwand zu realisieren.

Im Rahmen dieser Arbeit wurden die Zu- und Ablaufkonzentrationen der drei Benzotriazole in drei verschiedenen Kläranlagen über ein Jahr gemessen. 5-TTri zeigte guten biologischen Abbau mit einer mittleren Elimination von 75%, gefolgt von BTri mit 45% und schließlich 4-TTri mit 15%. Konzentrationsmessungen im Vorfluter vor und nach der jeweiligen Kläranlage bestätigten sie als Haupteintragsquelle für BTs ins Gewässer.

Um die gewonnenen Ergebnisse im Labor zu überprüfen wurden Abbauversuche durchgeführt. 30 mg L^{-1} BTri sowie 5-TTri konnten von Belebtschlammbiozönosen, die direkt aus der Kläranlage stammten, unter aeroben Bedingungen innerhalb von 21-49 beziehungsweise 2-7 Tagen biologisch abgebaut werden. Ein Abbau unter denitrifizierenden, sulfatreduzierenden oder anaeroben Bedingungen fand nicht statt, ebenso wenig wie ein biologischer Abbau von 4-TTri, unabhängig von den gewählten Bedingungen.

Durch Anpassung des Belebtschlamms an hohe BTri und 5-TTri Konzentrationen konnte die Abbauleistung wesentlich gesteigert werden, wodurch BTri innerhalb von sieben Tagen und 5-TTri innerhalb von vier Tagen vollständig eliminiert wurden. Zusätzlich angebotene Nährstoffe, v.a. Stickstoff, steigerten die biologische Abbauleistung zusätzlich, wohingegen zusätzlich angebotener Kohlenstoff nahezu keinen Einfluss auf die Eliminationsleistung zeigte.

Weiterführende Experimente zur Optimierung der Bedürfnisse der Belebtschlammbiozönose für den Abbau von 5-TTri durch gezielte Anpassung des Schlamms über mehrere Generationen hinweg erschufen eine für 5-TTri Abbau optimierte Biozönose mit einer Abbauleistung von 5,2 mg L^{-1} d^{-1}. Durch Zugabe eines Belebtschlammextrakts, um annähernd natürliche Nährstoffbedingung zu simulieren, zeigten auch nicht-angepasste Schlämme sehr hohe Abbauraten bis zu 5,0 mg L^{-1} d^{-1}. Eine ähnliche Verbesserung konnte durch die Zugabe von Stickstoff erzielt werden. Deshalb wird angenommen, dass der Abbau von Benzotriazolverbindungen mit der Spaltung des Benzolrings beginnt und deshalb Stickstoff, der nicht aus den Benzotriazolen gewonnen werden kann, den limitierenden Faktor darstellt.

Eine anschließende Charakterisierung der 5-TTri abbauenden Belebtschlammbiozönosen wurde mittels denaturierender Gradientengelelektrophorese (DGGE) und Next-Generation-Sequencing (Metagenom und Metatranskriptom Analyse) erreicht.

Die DGGE Resultate zeigten eine stark reduzierte Diversität mit vier dominanten Spezies: *Aminobacter* sp., *Flavobacterium* sp., *Hydrogenophaga* sp. und *Pseudomonas* sp..

Im Gegensatz dazu zeigte die Metagenomanalyse eine hohe Diversität und die Spezies *Mesorhizobium* spp., *Pseudomonas* spp., *Acidovorax* spp. und *Hydrogenophaga* spp. die größte Häufigkeit.

Die Analyse des Metatranskriptoms zur Bestimmung der metabolischen Aktivität der Organismen zeigte, dass die drei Spezies *Pseudomonas* spp., *Hydrogenophaga* spp. und *Acidovorax* spp. mit hoher Wahrscheinlichkeit am Abbau von 5-TTri beteiligt sind.

Ein weiterer Aspekt dieser Arbeit behandelte das biologische Abbauverhalten von SMX mittels neun verschiedener, aus Belebtschlamm isolierter, bakterieller Reinkulturen. Die Sequenzierung des 16S rRNA Gens lieferte fünf *Pseudomonas* spp., einen *Brevundimonas* sp., einen *Variovorax* sp. und zwei *Microbakterium* spp.. Inkubation dieser Organismen in Medien mit 10 mg L^{-1} SMX und leicht abbaubaren Energiequellen zeigte, dass je nach verwendetem Medium Abbauraten von bis zu 2,5 mg L^{-1} d-

[1] erreicht werden konnten. Zusätzlich wurde gezeigt, dass die gefundenen Kulturen fähig waren, SMX als alleinige Energie- und Nährstoffquelle zu nutzen.

LIST OF CONTENTS

1 CHAPTER 1 – INTRODUCTION 23
  1.1 Xenobiotic Organic Micropollutants – Issues Regarding the Aquatic Environment 25
  1.2 Xenobiotica – Challenges for Wastewater Treatment Systems 28
  1.3 Benzotriazole – Applications and Mechanism of Action 32
  1.4 Sulfamethoxazole – Application and Mechanism of Action 35
  1.5 General Objectives and Scope of this Work 37

2 CHAPTER 2 – SCREENING FOR MICROBIAL XENOBIOTICS' BIODEGRADATION BY UV-ABSORBANCE MEASUREMENTS 41
  2.1 Introduction... 43
  2.2 Materials and Methods 45
    2.2.1 Chemicals and Glassware 45
    2.2.2 Activated Sludge (AS).. 45
    2.2.3 Experimental Setup 45
    2.2.4 Analyses of SMX and BTs 48
  2.3 Results and Discussion 50
    2.3.1 Evaluation of UV-AM 50
    2.3.2 Parent Substances 50
    2.3.3 Screening for Transformation Products.................. 53
    2.3.4 Optimizing Media for UV-AM 54
    2.3.5 Screening ASCs for Xenobiotic Biodegradation 55
  2.4 Conclusions.. 62

3 CHAPTER 3 – MONITORING BENZOTRIAZOLES............65
  3.1 Introduction... 67
  3.2 Materials and Methods 68

- 3.2.1 Wastewater Treatment Plants and Receiving Water Bodies 68
- 3.2.2 Chemicals 71
- 3.2.3 Sample Preparation and Chemical Analyses of BTs 72
- 3.3 Results and Discussion 73
  - 3.3.1 Influent and Effluent Concentrations and Total Removal Efficiencies of the WWTPs 73
  - 3.3.2 Removal Efficiencies of the Wastewater Treatment Stages 77
  - 3.3.3 WWTPs Discharge into Receiving Rivers 80
- 3.4 Conclusions 84

4 CHAPTER 4 – BENZOTRIAZOLE BIODEGRADATION 87
- 4.1 Introduction 89
- 4.2 Materials and Methods 91
  - 4.2.1 Chemicals 91
  - 4.2.2 Wastewater Treatment Plants' Characteristics and Activated Sludge Sampling 92
  - 4.2.3 Experimental Setup 93
  - 4.2.4 Sample Preparation and Chemical Analyses of BTs 96
- 4.3 Results and Discussion 98
  - 4.3.1 Biodegradation of BTri, 4-TTri, and 5-TTri under Aerobic Conditions 98
  - 4.3.2 Biodegradation Behavior of BTri, 4- and 5-TTri under Low Redox Conditions 107
- 4.4 Conclusions 108

5 CHAPTER 5 – 5-TOLYLTRIAZOLE ACCLIMATION 111
- 5.1 Introduction 113
- 5.2 Materials and Methods 114
  - 5.2.1 Chemicals 114
  - 5.2.2 Experimental Setup for Biodegradation 115

5.2.3 Activated Sludge Inoculum 116

5.2.4 Acclimation Procedure 116

5.2.5 Specific Nutrient Supply 117

5.2.6 Detection of Biodegradation 118

5.3 Results and Discussion 118

5.3.1 Activated Sludge Acclimation at Low-Nutrient Conditions 118

5.3.2 5-TTri Biodegradation Enhancement by Nutrient Supply 123

5.4 Conclusion 130

6 CHAPTER 6 – METAGENOMIC ANALYSIS OF 5-TTRI BIODEGRADING ASC 133

6.1 Introduction 135

6.2 Materials and Methods 136

6.2.1 Chemicals and Reagents 136

6.2.2 Experimental Setup for Biodegradation 137

6.2.3 DNA and RNA Extraction for Next-Generation-Sequencing 139

6.2.4 Denaturing Gradient Gel Electrophoresis (DGGE) 139

6.2.5 Phylogenetic Analysis 140

6.2.6 Library Preparation and Next Generation Sequencing 141

6.2.7 Sequence-Data Processing 142

6.3 Results and Discussion 143

6.3.1 Biodegradation of 5-TTri under Aerobic Conditions and Different Media 143

6.3.2 DGGE Patterns of 5-TTri Biodegrading ASC 145

6.3.3 Metagenomic Analysis of 5-TTri Biodegrading ASC 152

6.3.4 Metatranscriptomic Analysis of 5-TTri Biodegrading ASC 156

6.4 Conclusions 158

7 CHAPTER 7 – SULFAMETHOXAZOLE BIODEGRADATION BY PURE CULTURES 161

7.1 Introduction 163

7.2 Materials and Methods 164

7.2.1 Chemicals and Glassware 164

7.2.2 Activated Sludge Sampling 165

7.2.3 Experimental Setup 165

7.2.4 Analyses of Sulfamethoxazole 167

7.2.5 Taxonomic and Phylogenetic Identification of Isolated Pure Cultures by 16S rRNA Gene Sequence Analysis 168

7.3 Results 169

7.3.1 SMX Biodegradation 169

7.3.2 Taxonomic and Phylogenetic Identification of Pure Cultures 171

7.3.3 SMX Biodegradation Studies with Pure Cultures ... 174

7.4 Discussion 180

7.5 Conclusions 184

8 CHAPTER 8 – CONCLUSIONS AND OUTLOOK 187

8.1 Conclusions 189

8.1.1 Efficient Screening for Biodegradation in Laboratory Setups 189

8.1.2 Benzotriazoles – Monitoring 190

8.1.3 Benzotriazoles - Biodegradation 192

8.1.4 Sulfamethoxazole – Identification of Pure, Biodegrading Bacterial Cultures 196

8.2 Outlook 198

9 REFERENCES 201

LIST OF TABLES

Table 1.1 – Physico-chemical properties of the three benzotriazole compounds and sulfamethoxazole 34

Table 2.1 – Media characterization, xenobiotic application setup and background absorbance at two different wavelengths used for monitoring xenobiotics' biodegradation. 46

Table 2.2 – Maximum absorbance (abs_{max}) wavelength of tested xenobiotic compounds and corresponding absorbance value for 10 mg L^{-1} 51

Table 2.3 – Comparison of UV-AM with LC-UV and GC-MS/MS measurements regarding time needed for analyzes and costs per sample 56

Table 2.4 - Biodegradation and time needed for absorbance decrease in the different ASC setups using different media, biomass concentration and initial xenobiotics´ concentrations. Shown are BTri, 5-TTri and SMX. 4-TTri was not biodegraded and is thus not shown. 61

Table 3.1 – Characteristics of wastewater treatment plants (WWTPs) analyzed and their receiving rivers 69

Table 3.2 – Mean influent and effluent concentrations with standard deviations of BTri, 5-TTri and 4-TTri over one year determined for the three WWTPs MBR-MH, CAS-E and CAS-M. 73

Table 3.3 – Mean BTs concentrations with standard deviations measured over one year in the receiving rivers upstream and downstream the corresponding WWTPs.......................82

Table 4.1 – Media compositions used for incubation. 93

Table 4.2 – Amounts of BTri, 4-TTri, and 5-TTri used for biodegradation experiments in the different setups. TTri consists of 4- and 5-TTri 40%/60% (w/w), respectively.......94

Table 4.3 – Biodegradation of 5-TTri under aerobic conditions with ASC CAS-M. Shown are initial 5-TTri concentration and time needed to remove 99.9% 5-TTri. 105

Table 4.4 – Xenobiotics´ biodegradation under low redox conditions. Shown are concentrations given in % from initial value (100%) over incubation time of 30 days for setups with activated sludge and 50 days for digester sludge. 107

Table 5.1 – Media compositions, carbon-nitrogen ratio, nutrient applications and experimental setups. 115

Table 5.2 – Biodegradation rates taken as the slope from the linear fitted lines of the ASC derived from the acclimation and nutrient experiments.. 120

Table 6.1 – Compositions, DOC to N ratios and applications of the used media for biodegradation experiments............... 137

Table 6.2 – Species found in the 5-TTri biodegrading ASC by sequence analysis of excised DGGE bands. Provided are sequence length and the similarity to the closest relative in the ENA database (http://www.ebi.ac.uk/ena/). 147

Table 7.1 – Initial (10 mg L^{-1}) and end concentrations of SMX accomplished with 12 biodegrading pure cultures gained out of 110 cultures. Taxonomic identification succeeded with BLAST (http://blast.ncbi.nlm.nih.gov/Blast.cgi). *duplicate organisms. All but SMX344 were discarded. ... 170

Table 7.2 – Biodegradation rates of the cultures able to biodegrade SMX. Isolation was performed from an SMX-acclimated ASC, followed by identification with 16S rRNA sequencing. ENA accession numbers and species names are provided. * calculated from duplicate experiments (n=2). Standard deviations between duplicate setups were below 1% and are not shown. 174

LIST OF FIGURES

Figure 1.1 A, B and C – Structures of A) benzotriazole, B) 4-methyl-benzotriazole and C) 5-methyl-benzotriazole 32

Figure 1.2 – Proposed structure of the benzotriazole surface layer on copper [57]. 33

Figure 1.3 A and B – Structures of A) sulfamethoxazole (SMX) and B) trimethoprim 35

Figure 1.4 A and B – Pathway of tetrahydrofolate synthesis in bacterial cells. 37

Figure 2.1 A and B – A) UV-absorbance spectra of tested xenobiotic compounds in pure water at 10 mg L^{-1}. 52

Figure 2.2 - Effect of used media on UV-absorbance. 53

Figure 2.3 - Correlation of UV-AM results with LC-UV and GC-MS/MS analyses. 57

Figure 2.4 A and B - Time series of SMX biodegradation with activated sludge communities in MSM-CN (A) and MSM (B) as detected by UV-AM. 60

Figure 3.1 – Flow charts of the monitored WWTPs MBR-MH, CAS-E and CAS-M including the different treatment stages and sampling points (SP). 70

Figure 3.2 – Influent and effluent concentrations of the three BTs BTri, 4-TTri, and 5-TTri and the daily influent volume of the three WWTPs: A) MBR-MH, B) CAS-E, and C) CAS-M. 75

Figure 3.3 – Mean BTri, 4-TTri and 5-TTri removal efficiencies with standard deviation (n=12) over one year of the three WWTPs, MBR-MH, CAS-E and CAS-M. 77

Figure 3.4 – Removal efficiencies per sampled treatment stage of 5-TTri, BTri and 4-TTri over one year analyzed for the WWTPs A) MBR-MH, B) CAS-E and C) CAS-M. 78

Figure 3.5 – Concentrations of BTri, 4-TTri and 5-TTri upstream and downstream the WWTPs over one year.........................83

Figure 4.1 – Chemical structures and retention times, according 4.2.4, of the three different benzotriazole compounds and 5,6-methyl-benzotriazole used as internal standard............92

Figure 4.2 – Aerobic BTri biodegradation patterns within the three ASCs. 102

Figure 4.3 – BTri biodegradation with ASC MBR-MH (A), CAS-E (B) and CAS-M (C) in four different media (Table 4.1)...103

Figure 4.4 – 5-TTri biodegradation with ASC CAS-M and five different media (Table 4.1)...106

Figure 5.1 – 5-TTri removal monitored by UV-absorbance measurements...121

Figure 5.2 – 5-TTri removal in sludge supernatant supplied setups monitored by UV-absorbance measurements. 124

Figure 5.3 A, B, C – 5-TTri removal monitored by UV-absorbance measurements. 129

Figure 6.1 – 5-TTri biodegradation patterns observed with UV-AM in three different media (Table 6.1) inoculated with ASC from generation one and eight. DNA analyses with DGGE were performed at day 7, 9 and 22, DNA and RNA analyses with next-generation-sequencing at day 9..........144

Figure 6.2 – DGGE band profiles of the ASC from generation one and eight after seven, nine and 22 days of incubation (see Figure 6.1) in two different media (Table 6.1)............146

Figure 6.3 – Magnification of DGGE band profiles of ASC amplified with primer set 27f_GC-517r. Visible gradient ranging from 70 to 75%. 146

Figure 6.4 – DGGE band profiles of the ASC, amplified with primer set 341f_GC-907r, from generation one and eight after seven, nine and 22 days of incubation and two different media (Table 6.1)...150

Figure 6.5 – DNA and RNA analyses of the *de novo* assembly of the sequences mapped against the complete NCBI database. 153

Figure 6.6 – Principal component analysis of the DNA (○) and RNA (●) extracted from the 5-TTri biodegrading setups at day nine. 155

Figure 7.1 – Absorbance patterns of isolated pure cultures in MSM-CN. 171

Figure 7.2 A and B – Maximum likelihood-based trees reflecting the phylogeny and diversity of the isolated nine species capable of SMX biodegradation. 172

Figure 7.3 A and B – A) Aerobic SMX biodegradation patterns of pure cultures in R2A-UV media 176

Figure 7.4 A, B, C and D – A, B) Aerobic SMX biodegradation patterns of pure cultures in MSM-CN 177

Figure 7.5 A, B, C and D – A, B) Aerobic SMX biodegradation patterns of pure cultures in MSM media 178

LIST OF ABBREVIATIONS

µ	micro
µg	microgram
4-TTri	4-tolyltriazole = 4-methyl-benzotriazole
5-TTri	5-tolytriazole = 5-methyl-benzotriazole
abs	absorbance
AS	activated sludge
ASC	activated sludge community
BLAST	Basic Local Alignment Search Tool
BOD	biological oxygen demand
BTri	1-H-benzotriazole
BTs	all three benzotriazoles together
C	concentration
CAS	conventional activated sludge
CAS-No.	chemical abstracts service number
DGGE	denaturing gradient-gel electrophoresis
DOC	dissolved organic carbon
ENA	European Nucleotide Archive
F/M ration	food to microorganism ratio
g	centrifugal force
GC	gas chromatography
HRT	hydraulic retention time
L	liter
LC	liquid chromatography
m	milli
max	maximum
MBR	membrane bioreactor
MIC	minimum inhibitory concentration
MLSS	mixed liquor suspended solids
MS	mass spectrometry
MSM	mineral salt medium
MSM-C	mineral salt medium supplied with carbon

MSM-CN	mineral salt medium supplied with carbon/nitrogen
MSM-N	mineral salt medium supplied with nitrogen
MSM-PE	mineral salt medium supplied with peptone
MSM-SS	mineral salt medium supplied with sludge supernatant
n	nano
ng	nanogram
PBS	phosphate buffered saline
PCR	polymerase chain reaction
PE	population equivalents
R^2	R squared (coefficient for determination)
R2A	Reasoner's 2A medium
R2A-UV	R2A modified for UV-AM
SMX	sulfamethoxazole
SP	sampling point
sp.	species
spp.	species pluralis
SS	sludge supernatant
TTri	tolyltriazole
UV-AM	ultraviolet-absorbance measurements
v/v	volume per volume
w/w	weight per weight
WWTP	wastewater treatment plant

Chapter 1

Introduction

1.1 XENOBIOTIC ORGANIC MICROPOLLUTANTS – ISSUES REGARDING THE AQUATIC ENVIRONMENT

Xeno|biotica (gr. Βιοτικός alive, able to live) n pl: *1. Compounds inducing defense reactions in the human body (antigens, toxins...) 2. Substances that are foreign for an ecosystem, e.g. pollutants; compare ecology.* (adapted from [1])

Micro|pollutants: *synthetic and natural trace contaminants being present in aquatic systems at low to very low concentrations ranging from pg per liter to µg per liter.* (adapted from [2])

Chemical and pharmaceutical industries produce a large variety of products that are omnipresent in our everyday life. These products are necessary to maintain the high living standard that our society is used to and they are implicitly required for communication technologies, food supply, mobility and safety issues. On the other hand, many of these xenobiotic compounds, i.e. substances that do not naturally occur in ecosystems, contaminate aquatic systems as many of them are persistent against biodegradation and/or are only poorly removed during wastewater treatment. With the development of sensitive and easily applicable analytical techniques to monitor the concentrations of various xenobiotica it has been realized that almost all aquatic systems, including ground water, are prone to contamination. Although many of these substances are detectable only in ng L^{-1} concentrations and are thus entitled micropollutants, they may pose a potential risk on human health and cause detrimental effects on aquatic ecosystems [3]. Water environments often have a direct and immediate contact with human life as water is almost used ubiquitously. Furthermore, water bodies form a unit allowing xenobiotica to be transported to every part of the world [4]. Mostly these compounds, especially when related to personal applications like fragrances, hygiene products or health-related compounds, enter the water cycle in excessive amounts as they are intensively applied [5, 6]. Additionally, pesticides and antibiotics that are used in veterinary applications enter the water cycle due to their use in agriculture and farming [7, 8]. In contrast to the latter, that enter aquatic systems without treatment by direct runoff, are those that are discharged with wastewater treatment plant (WWTP) effluents. Many xenobiotica are biologically stable and/or polar compounds and thus are only partly removed during wastewater treatment [9]. Personal care products, flame retardants, surfactants and pharmaceuticals are among these products that exhibit a low biodegradability and thus weak retention in WWTPs [10, 11]. These compounds, even if they are discharged in the lower ng L^{-1} range only, comprise the risk to accumulate in aquatic systems as their self-purification potential might not be sufficient enough to completely eliminate these substances [12, 13]. Therefore, pharmaceuticals have already been detected in drinking water in concentrations up to the lower µg L^{-1} range [14] with highest concentrations of 1.3 µg L^{-1} for Ibuprofen [15]. Even in such low concentrations, some compounds like human pharmaceuticals, especially antibiotics, are known to exhibit adverse effects in aquatic ecosystems [16, 17]. Evidence has shown that even in sub-inhibitory level concentrations antibiotics might exert an impact on microbial communities, e.g. by influencing transcription in bacteria [18], but long-term effects on exposure to low concentrations of antibiotics still remain largely unknown. Additionally, only few information is available on the potential expansion of antibiotic resistances arising from the permanent exposure of aquatic organisms towards antibiotics [19] although many studies address the detection of antibiotic resistance genes and antibiotic-resistant organisms in aquatic systems, ground- and drinking water [20, 21]. However, it was reported that continuous exposure of microbial activated sludge communities (ASC) to low concentrations of 100 µg L^{-1} erythromycin significantly changed the community structure [22]. If long term exposure to low levels of xenobiotica also has adverse effects on human health still remains unknown due to a lack of proper studies [23]. Nevertheless, all xenobiotica are designed to show a specific effect and also low concentrations might be enough to cause undesired side-effects as a study with fathead minnows showed [24]. In this study fathead minnows (*Pimephales promelas*) were chronically exposed to low concentrations of 5 ng·L^{-1} 17α-ethynylestradiol that already led to feminization of males and ultimately to a near extinction of this species. This is of special importance as the applied concentrations were by far lower than many of the already detected concentrations of xenobiotic compounds in aquatic systems. This example clearly demonstrated that xenobiotic micropollutants, even in lowest concentrations, can influence aquatic biological systems and should thus be avoided on the one hand by reducing their application and, on the other hand, by improving wastewater treatment systems removal of persistent organic pollutants.

1.2   XENOBIOTICA – CHALLENGES FOR WASTEWATER TREATMENT SYSTEMS

As many xenobiotica are present only in low to very low concentrations, i.e. µg to ng L^{-1} range, conventional wastewater treatment systems are often incapable of completely removing such compounds [25]. Thus, such micropollutants are often detected in the effluent waters of WWTPs [26, 27] by newly developed and/or improved chemical analysis, i.e. LC and GC methods [28], for quantitative measurements of very low concentrations. In contrast, qualitative analyzes are mainly performed by biological test systems to show the effects of the released micropollutants on ecosystems. Sensitive and specific biomarkers as the egg yolk protein vitellogenin [29] or a bioassay using transgenic medaka embryos [30] for measuring the estrogenic activity *in vivo* were established for specifically screening for changes in the applied organisms. Another vital issue of environmental impacts is the presence of multicomponent mixtures that may lead to combined effects as shown for fathead minnows regarding estrogenic response and reproductive performance [31]. These issues show that modern wastewater treatment not only requires removal of carbon, nitrogen and phosphorus compounds but should also focus on the elimination of as many micropollutants as possible to reduce their concentrations. WWTPs constitute major point sources of such pollutants in their receiving rivers. Increasing their removal efficiency would thus significantly decrease xenobiotica concentrations downstream WWTPs [32]. Very many treatment plants operate variants, e.g. one-stage, two-stage, intermittent systems, of the well-known conventional activated sludge process (CAS) but also the quite new and innovative technology of the membrane bio-reactor (MBR) is gaining popularity [33]. AS treatment is limited as its removal mainly depends on biodegradation and sorption onto biomass. Often it is operated with relatively short hydraulic retention times (HRT, one to six hours) and lower sludge ages (up to 12 days) compared to MBR systems with HRTs of around 10 to 20 hours and sludge ages up to 40 days. These limitations favor MBR over CAS systems for the removal of trace organic pollutants as was shown by several studies [34, 35]. The main factors for a better removal of such compounds in MBR systems might be the possibility to achieve a high sludge age that allows for the enrichment of slow growing bacteria and favors a high microbial diversity [36]. Therefore, genetic mutations can be established and microorganisms' acclimation to the assimilation of persistent organic compounds increases [37]. In addition, it was found that enzymatic production and thus the total enzymatic activity in the system increases when the flocks are of smaller size (mean 4 µm in MBR to 20 µm in CAS [38]) due to an increase in their specific surface area [39]. Thus, mass-transfer conditions in the MBR are improved. Moreover, the presence of many planktonic bacteria increases the cells-specific metabolic processes as their higher specific surface implies shorter distances to overcome to reach possible substrates. This phenomenon could possibly lead to an enhanced xenobiotica removal in MBR systems [40, 41].

However, also MBR systems, although showing mostly better xenobiotica removal than CAS, still release micropollutants with their effluents. Thus, post-treatment methods that might be retrofitted to existing WWTPs as a forth treatment stage like activated carbon or ozone treatment, advanced oxidation technologies or sonolysis [42] are considered as they seem to be an efficient method to remove such substances [43]. These techniques, capable of removing pharmaceuticals, corrosion inhibitors and other xenobiotica [44, 45] are also prone to some severe disadvantages: they often require higher maintenance, have a higher energy consumption and thus are cost-intensive and, due to their unspecific reaction mechanisms, tend to form highly reactive products of unknown behavior or toxicity [46]. Research has to assess this risk by monitoring the effluents of advanced water treatment plants by bioassays [47]. However, further research is vital to exclude that micropollutants transformation products, whether originating from biodegradation or from post-treatment, have an even higher detrimental impact on the receiving aquatic system than the original compound. Studies on ozonation transformation products formed during the oxidation of the beta-blocker propranolol already showed that very reactive species like aldehydes are likely to be formed [48] as well as inorganic toxic compounds like bromates [49]. To conclude, as long as these concerns remain unclear advanced oxidation processes, apart from enabling an almost complete removal of various compounds, should be used with caution for polishing up WWTPs effluents. It might well be possible that unknown toxic reaction products are released into the environment. In the worst case they could show an even higher negative impact on aquatic communities than the parent compound or infiltrate ground and ultimately drinking water.

In addition, biodegradation, apart from sorption onto biomass and the above mentioned techniques, with activated sludge communities still represents a powerful approach to efficiently reduce xenobiotica effluent concentrations. Very many studies elucidated that xenobiotica are sometimes well biodegraded during activated sludge treatment and thus do not require further polishing and sometimes are very persistent. Additionally, the biological removal varies strongly from compound to compound and does not allow correlation to the compound structure [26, 32]. Among these are pharmaceuticals like ibuprofen and naproxen that are well biodegraded whereas others like carbamazepine and sulfamethoxazole show only a low removal due to biodegradation [32]. Sulfamethoxazole biodegradation by AS in laboratory setups is already described by several studies [19, 50, 51]. However, knowledge about optimal biodegradation conditions and biodegradation patterns with pure cultures is still very rare [52].

Not only pharmaceuticals but also compounds, generally referred to as polar persistent organic pollutants (PPOPs) [17] show a weak biodegradability. The latter include, among others, personal care products and industrially relevant substances like the corrosion inhibitors benzotriazole. Benzotriazoles, especially 1-H-benzotriazole, 4- and 5-tolyltriazole, exhibit, like sulfamethoxazole, a weak biological removal during wastewater treatment and are thus discharged in concentrations up to µg L^{-1} [53]. Unlike with sulfamethoxazole, benzotriazoles' biodegradation still lacks specific laboratory studies to evaluate AS communities biodegradation patterns and capacities. For these reasons, sulfamethoxazole and the benzotriazoles were chosen for evaluation of their biodegradation likeliness under different nutrimental conditions with AS communities as well as pure cultures.

1.3 BENZOTRIAZOLE – APPLICATIONS AND MECHANISM OF ACTION

Benzotriazoles (BTs) are a class of substances that are mainly used for corrosion protection of various metal surfaces, especially copper [54]. They are entitled as "emerging contaminants", i.e. pollutants that occurred in the last 20 years and are being found in significant concentrations and abundance [55].

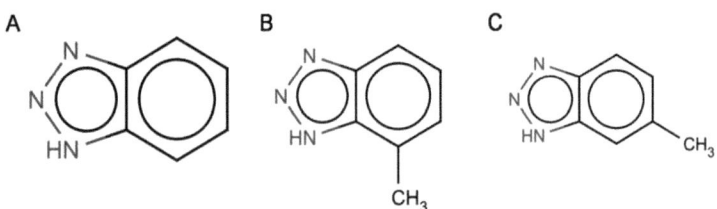

Figure 1.1 A, B and C – Structures of A) benzotriazole, B) 4-methyl-benzotriazole and C) 5-methyl-benzotriazole

While benzotriazole (Figure 1.1 A) is preferentially used as additive to dishwashing detergents, the two methyl-benzotriazoles 4- and 5-tolyltriazole (Figure 1.1 B and C), summarized as tolyltriazole, are industrially applied and added to braking fluids and aircraft deicing formulations [56]. These compounds are very efficient as only very low concentrations are necessary to successfully prevent corrosion [58].

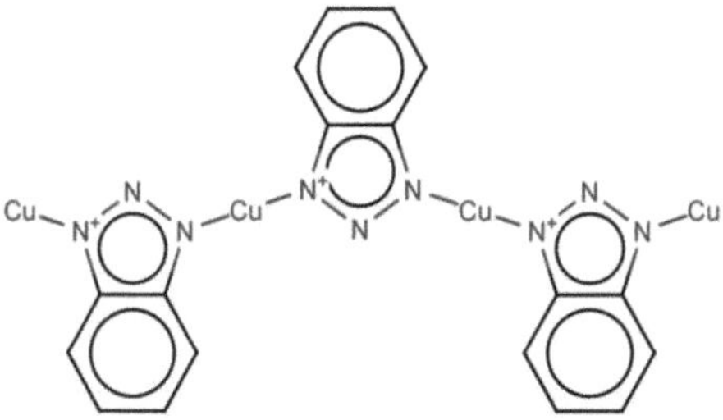

Figure 1.2 – Proposed structure of the benzotriazole surface layer on copper [57].

The benzotriazole particles form a thin but stable organic barrier layer on metal, especially copper, surfaces and thus efficiently prevent corrosion [59]. However, the exact mechanisms that lead to this layer are still object of discussions. It seems possible that benzotriazoles form differently structured surface layers with different types of metal [60]. The mechanisms that connects benzotriazoles to the metal surfaces are also still not completely understood and object of further research [61]. One proposed structure (Figure 1.2) shows the formation of a Cu–N bond that is explained by overlapping of the Cu sp-hybrid atomic orbital with the N-hybridized lone-pair atomic orbitals [57]. Apart from the benzotriazoles good corrosion protection are other factors that favors their use in a wide range of applications: they are very cheap with around 5 € kg^{-1} (personal communication [62]), are well water soluble (see Table 1.1) and very easy to use as they can be applied as aqueous solution.

Table 1.1 – Physico-chemical properties of the three benzotriazole compounds and sulfamethoxazole

compound	CAS-No.	molecular formula	MW [g mol^{-1}]	water solubility (20 °C)	logP	appearance
1H-benzotriazole (BTri)	95-14-7	$C_6H_5N_3$	119.12	19.0 g L^{-1}	1.26	slightly yellow granulate
5-methyl-benzotriazole (5-TTri)	136-85-6	$C_7H_7N_3$	133.15	6.0 g L^{-1}	1.78	cream to beige powder
4-methyl-benzotriazole (4-TTri)	2987 8-31-7	$C_7H_7N_3$	133.15	0.2 g L^{-1}	1.78	yellow to brownish powder
Sulfamethoxazole (SMX)	723-46-6	$C_{10}H_{11}N_3O_3S$	253.28	0.4 g L^{-1}	0.79	white powder

These factors make the three benzotriazoles inevitable for various applications such as additives to dishwashing detergents [63], cooling and metal working fluids [64], UV-filters and sunscreen agents [65, 66], additive to anti-freeze fluids [67], as antifoggant in photography applications [68], for conservation purposes of antiques [69]. Moreover, they are contained in cosmetics, skin creams, and body lotions as stabilizers [65]. Their diverse application and weak retention during wastewater treatment result in their almost omnipresent occurrence in all larger aquatic systems across Europe [53] and implies the need for research on how to improve their elimination during wastewater treatment.

1.4 SULFAMETHOXAZOLE – APPLICATION AND MECHANISM OF ACTION

The discovery of penicillin in 1929 by Alexander Fleming [70] completely changed our view on infectious diseases that suddenly became curable, contributed to new insights into microbiological biology and shifted patients' expectations and structures of drug companies [71]. After the first euphoria the development of new antibiotics ceased and drug companies shifted focus onto more profitable sectors like the development of drugs for cancer and heart diseases [72]. For that reason almost the same antibiotics with similar modes of action have been used over the last decades [73] that allowed for the development of various bacterial resistances against almost all classes of antibiotics as well as multi-drug-resistant bacteria [74].

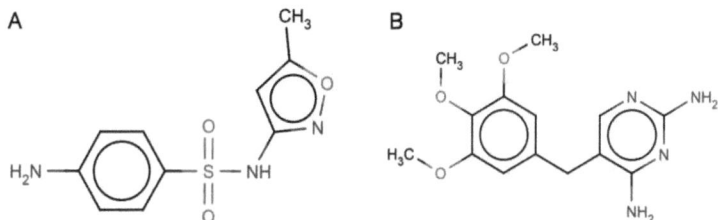

Figure 1.3 A and B – Structures of A) sulfamethoxazole (SMX) and B) trimethoprim

Therefore, the sulfonamide antibiotic sulfamethoxazole (Figure 1.3 A, Table 1.1), a commonly applied antibiotic to treat human urinary tract infections or pneumonia [75], is no longer potent enough and thus only applied together with trimethoprim (Figure 1.3 B). This combination, commonly known as co-trimoxazole (SXT, TMP-SMX, TMP-SMZ or TMP-sulfa), is preliminarily used to treat a large variety of bacterial infections as this combination was more effective than each of its components [76]. Sulfamethoxazole is a structural analog and competitive antagonist of para-aminobenzoic acid (PABA) and prevents the formation of folic acid that is implicitly required for DNA synthesis (Figure 1.4 B). Trimethoprim, in contrast, binds the dihydrofolate reductase and inhibits the reduction of dihydrofolic acid to tetrahydrofolic acid (Figure 1.4 A). Both antibiotics inhibit the formation of folic acid and thus are characterized as bacteriostatic, i.e. they do not kill the bacteria (bactericide) but prevent further growth. Bacterial cells must synthesize folic acid as they are not able, unlike humans, to up take that important molecule in form of vitamin B_9 [77].

Due to SMX' great and widespread application to treat human as well as animal infections worldwide [78], sulfonamide resistances are omnipresent among various bacterial species [79]. The mechanisms to cause sulfamethoxazole resistance in bacteria leads to the formation of an altered dihydropteroate synthase (Figure 1.4) with a reduced affinity for sulfamethoxazole but an increased affinity for PABA. The expression of that altered enzyme is based on two different mechanisms: A) By expression of plasmid-coded genes named Sul1, Sul2 and Sul3 and B) by mutation of the folP gene on the chromosomal DNA [80]. While the dissemination of the folP gene is limited to vertical gene transfer, the plasmid coded Sul genes are easily spread among different bacterial species by horizontal gene transfer and can induce a fast adaption of various bacteria to sulfamethoxazole as was recently shown for soil microbial communities and the Sul1 gene [81].

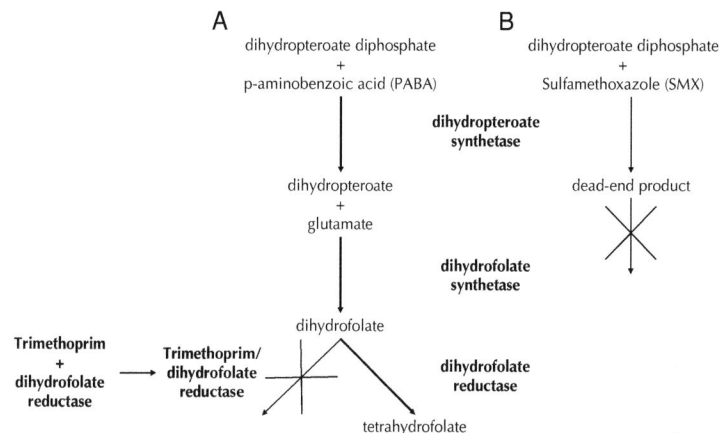

Figure 1.4 A and B – Pathway of tetrahydrofolate synthesis in bacterial cells. A) Valid pathway (bold arrows) and blocked pathway by trimethoprim and B) blocked pathway by sulfamethoxazole.

These general concerns about antibiotic resistances that might spread among different bacteria due to the low retention of sulfamethoxazole during wastewater treatment imply the need for further research on sulfamethoxazole biodegradation.

1.5 GENERAL OBJECTIVES AND SCOPE OF THIS WORK

Previous studies on benzotriazoles mainly focused on monitoring and comparing different wastewater treatment systems for their effectiveness in removing BTri, 4-TTri and 5-TTri, whereas in many reports 4-TTri and 5-TTri were not monitored separately but only as the mixture tolyltriazole (TTri) [17, 53]. These monitoring approaches in wastewater treatment plants only compared the influent and effluent BTs concentrations without taking into account the contribution of the different treatment stages [26, 82]. However, knowledge about the contribution of the treatment stage could significantly improve the removal of BTs as it would allow adapting the treatment system for an optimized BTs elimination. Furthermore, this knowledge would allow to identify the main elimination process, e.g. biodegradation, sorption, photolysis, that occurs in the different stages.

Further studies reported the possible sources of BTs in the aquatic systems as there are dishwashing detergents [83], industry [84, 85], food processing companies as they use BTs in their cooling systems [86] and BTs as additives to deicing/antifreeze formulations [87, 88]. Additionally, recent laboratory studies, conducted with AS and focusing on the biodegradation of BTri and 5-TTri, showed that these compounds behave differently during biodegradation under aerobic, anoxic and anaerobic conditions [89, 90]. However, investigations on the removal of 4-TTri are still missing and so do studies taking into account the applied nutrient conditions or differences arising from the sludge origin.

Sulfamethoxazole, in contrast to the BTs, is a well-characterized antibiotic that shows a good removal due to biodegradation during wastewater treatment [51] as well as in laboratory experiments conducted with activated sludge [50]. Although the mode of action and the resistances in bacteria for SMX are well understood, knowledge on biodegrading species, biodegradation pathways, and optimization of its removal, especially during wastewater treatment, still lacks a lot of understanding. However, this knowledge would permit to specifically improve SMX removal or

Chapter 2

Monitoring Microbial Xenobiotics' Biodegradation via Rapid and Inexpensive Microplate UV-Absorbance Measurements (UV-AM)

Evaluation of xenobiotics biodegradation potential, shown here for benzotriazoles and sulfamethoxazole by microbial communities and/or pure cultures normally requires time intensive and money consuming LC/GC methods that are, in case of laboratory setups not always needed. When using high concentrations in laboratory setups, a simple UV-absorbance measurement (UV-AM) that was developed and validated, can be applied. A large number of setups can be screened with almost no preparation and significantly less time and money compared to LC/GC methods being required. Evaluation was performed by comparing its measured values to LC-UV and GC-MS/MS results. Furthermore its application for monitoring and screening unknown activated sludge communities (ASC) and mixed pure cultures was tested and approved to detect biodegradation of benzotriazole (BTri), 4- and 5-methyl-benzotriazole (4-TTri, 5-TTri) as well as SMX. In laboratory setups, xenobiotics concentrations above 1.0 mg L^{-1} without any enrichment or preparation could be detected after optimization of the method.

2.1 INTRODUCTION

Evaluation and monitoring of the biodegradation potential of different activated sludge communities (ASC) as well as other microbial systems is often time consuming and money intensive. Mostly techniques like LC-MS/MS or GC-MS/MS are applied for determination of concentrations of various compounds and, to a lesser extent, also LC-UV. When very low concentrations (µg L^{-1} or ng L^{-1}) need to be measured these techniques are the only option, but in many laboratory biodegradation setups under standardized conditions much higher concentrations in the mg L^{-1} range are used. Furthermore, for pre-testing biodegradation potentials in a large number of setups, LC/GC methods require too much time and are often not necessary as knowing exact concentration values are not needed as it is sufficient enough to know whether biodegradation occurs or not. Therefore, a rapid, easy to use and inexpensive technique is required to screen a large number of different setups for their biodegradation potential towards different xenobiotica. In a research project benzotriazoles (BTs) and the antibiotic sulfamethoxazole (SMX) were used as xenobiotica to evaluate their biodegradation pattern in laboratory setups.

These xenobiotic compounds are polar micropollutants with a wide spectrum of use. BTs are extensively used as corrosion inhibitors [65] while SMX, although mainly applied in veterinary medicine, is still one of the most commonly used antibiotics worldwide [91, 92]. Both compounds show high water solubility, an ubiquitary occurrence in almost all water bodies and an incomplete biological removal [16, 84, 90, 93-95]. Former studies already reported that wastewater treatment plants (WWTP) constitute one major point source for these compounds to be released into the aquatic environment [96-99]. Therefore, biodegradation studies, performed under specific laboratory conditions to exclude abiotic processes, are implicitly required to gain information about the biological removal potential of ASCs as they are one way to reduce the input of these compounds into aquatic environmental systems [34, 100, 101]. Laboratory experiments already showed a completely different removal behavior of benzotriazole (BTri), 4- and 5-methyl-benzotriazole (4-TTri, 5-TTri) [102] but biodegradation conditions remain rather unclear. SMX, in contrast, showed sometimes an almost complete removal in lab-scale setups inoculated with ASCs and/or mixed cultures under different conditions tested [50, 103]. Furthermore only little information is available on individual organisms being capable of SMX biodegradation as well as biodegradation potential under different redox and nutrient conditions [52, 104, 105].

To address the need for a rapid screening, this study provides a simple and inexpensive method to evaluate the potential of ASCs, mixed bacterial pure culture communities as well as single pure culture bacteria to biodegrade BTs and SMX. A test system for biodegradation detection that requires almost no preparation, uses simple UV absorbance measurements (UV-AM), and can be performed in microplate setups was developed and evaluated by comparing its results with LC-UV and GC-MS/MS. That system allows screening a large number of setups, minimizing laboratory costs and experimental time.

2.2 MATERIALS AND METHODS

2.2.1 Chemicals and Glassware

1-H-benzotriazole (BTri) and tolyltriazole (TTri), consisting of 4-tolyltriazole (4-TTri) and 5-tolyltriazole (5-TTri) (40%/60%, w/w), were provided by Cimachem GmbH (Kirchheimbolanden, Germany). 5-TTri, 4-TTri and sulfamethoxazole (SMX) were obtained from Sigma Aldrich (Steinheim, Germany) as well as sodium carbonate solution and toluol. All other chemicals were purchased from Merck KGaA (Darmstadt, Germany). High-purity water used for media, solutions, buffers, and analyses was prepared by a Milli-Q system (Millipore, Billerica, MA, USA). All used glassware was from Schott AG (Mainz, Germany).

2.2.2 Activated Sludge (AS)

AS was taken from biological stage 1 of a 2-stage municipal conventional activated sludge-plant (CAS-M) that is treating 1 million population' equivalents. The influent consists of municipal as well as up to 50% industrial wastewater. 500 mL AS were collected for reactor inoculation in pre-cleaned 1L glass bottles, stored at 4°C and used within 24h.

2.2.3 Experimental Setup

2.2.3.1 Inocula and Media

Biodegradation capabilities of AS towards BTs and SMX were tested in separate setups. Fresh AS was centrifuged (10 min, 4000 g), the supernatant discarded and the remaining biomass washed with 1xPBS-Buffer (NaCl (8.0 g L^{-1}), KCl (0.2 g L^{-1}), Na_2HPO_4 (2.7

g L^{-1}), KH_2PO_4 (0.2 g L^{-1})). The procedure was repeated twice to remove wastewater residues. 20 mL of one of the three liquid media (Table 2.1) were inoculated to an initial concentration of 0.5 g L^{-1} MLSS (mixed liquor suspended solids), 2.0 g L^{-1} MLSS and 3.5 g L^{-1} MLSS, respectively. These media, different in nutrient composition, were used for detection of BTs and SMX biodegradation by UV-absorbance measurements (Table 2.1). Media were adapted to fit the needs of UV-AM and determine variations in BTs and SMX biodegradation concerning availability of nutrients, degradation rate and test reliability of UV-AM.

Table 2.1 – Media characterization, xenobiotic application setup and background absorbance at two different wavelengths used for monitoring xenobiotics' biodegradation.

media	components [g L^{-1}]	pH	xenobiotics' application	absorbance [nm] 257	262
MSM (mineral salt media)	KH_2PO_4 (0.08), K_2HPO_4 (0.2), Na_2HPO_4 (0.3), $MgSO_4$*7 H_2O (0.02), $CaCl_2$*2 H_2O (0.04), $FeCl_3$*6 H_2O (0.003), Hoagland trace elements (0.1 mL L^{-1})	7.4	xenobiotics as sole C and N source	0.07	0.07
MSM-CN (+ C and N)	as MSM including: sodium acetate (0.5) and NH_4NO_3 (0.01)	7.4	xenobiotics' co-metabolism	0.08	0.09
R2A-UV (for UV-AM)	casein peptone (1.0), glucose (0.5), sodium acetate trihydrate (0.5), KH_2PO_4 (0.3), soluble starch (0.3), Hoagland trace elements (0.1 mL L^{-1})	7.2	foster growth of many organisms, xenobiotics' co-metabolism	0.53	0.56

2.2.3.2 Batch Tests

Reactors, inoculated to one of the mentioned MLSS concentrations, were set up in parallel. Each media was spiked with

BTri, 4-TTri, 5-TTri and SMX at concentrations of 1, 10, 25, 50 and 100 mg L^{-1}. Experiments were performed with R2A-UV media for optimal growth conditions, MSM-CN for testing the need of carbon/nitrogen and MSM for highly selective growth conditions due to nutrient depletion. Concentrations of BTs/SMX were varied to ensure acclimation and increased selectivity. In total 630 experiments were set up (7 compounds, 5 concentrations, 3 media, 3 different MLSS concentrations, each in 2 parallels).

All experiments were carried out in the dark to avoid photolysis. Aerobic conditions were ensured by shaking the reactors at 150 rpm on an orbital shaker. Temperature was kept constant at 20°C (± 2°C), pH controlled to be in the range of 7 to 8.

All experiments were performed until absorbance reached a stable and distinctly lower value compared to the initial one. In case no change in absorbance was detected, the experiment was stopped after a maximum of 56 days.

2.2.3.3 Sampling and Sample Pre-Treatment

Sampling of the reactors and detection of biodegradation by UV-AM was carried out once a day. 1.0 mL supernatant was taken from each reactor after 30 min sedimentation to reduce biomass withdrawal and media turbidity. 200 µL were used for UV-AM, 800 µL for LC-UV (SMX) or GC-MS/MS (BTs) analyses. Samples for UV-AM were analyzed the same day, remaining samples were stored frozen at -20°C until analysis.

2.2.4 Analyses of SMX and BTs

2.2.4.1 UV-AM Microplate Test System

200 µL were taken from the reactors and directly used for measurement. Centrifugation (10 min, 8,000 g, 20°C) and/or filtration (0.25 µm, PTFA syringe filter) were applied in case of high cell density (high turbidity) to remove cells and/or cellular debris. Analyses were performed in 96 well plates with a UV-permeable bottom foil (lumox® multiwell, Sarstedt AG, Nürnbrecht, Germany) using an automated plate reader (EnSpire® Multimode Plate Reader, Perkin Elmer, Rodgau, Germany). UV spectra of all used compounds from 230 to 330 nm at 10 mg L^{-1} in pure water and the three used media were recorded and absorbance maxima of the substances determined. Calibration was carried out for all tested compounds with 0.1, 1.0, 5.0, 10.0 and 20.0 mg L^{-1} in pure water and used media. Background absorbance of used 96 well UV-plates filled with 200 µL high-purity water (=working volume) was 0.07– 0.09 units. Linear absorbance, due to the internal calibration of the reader itself, was ensured when the absorbance lay within a range of 0.2 to 1.0 absorbance units. For each measurement a blank (media and xenobiotic, without biomass) was measured to detect changes over time as well as a zero blank (only media) to detect background absorbance.

2.2.4.2 GC-MS/MS and LC-UV Analyses

Samples from the reactors were centrifuged (10 min, 8,000 g, 20°C), filtered through an 0.45 µm PTFE filter, filled into sterile glass flasks for LC/GC, and stored in the dark at -20°C upon analysis.

LC-UV measurements were performed with a Dionex 3000 series HPLC system (Dionex, Idstein, Germany), equipped with an autosampler. DAD scanning from 200 to 600 nm was applied to detect and quantify SMX. Limit of quantification and limit of detection were 0.1 mg L^{-1} and 0.03 mg L^{-1}, respectively.

Chromatographic separation was performed on a Nucleosil 120 - 3 C18 column (250 mm x 3.0 mm i.d., 3 µm particle size) from Macherey Nagel (Düren, Germany). Column temperature was 25 °C. The mobile phases were acetonitrile (AN) and water (pH 2.5 using phosphoric acid). The gradient used was 0-5 min, 7% AN; 5-18 min, 7-30% AN; 18-30 min, 30% AN; 30-35 min, 7% AN. The solvent flow rate was 0.6 mL min^{-1}. The column was allowed to equilibrate for 5 min between injections.

For GC-MS/MS measurements, chemical preparation and determination of the BTs' concentration in the liquid samples were performed according to Liu et al. (2011), with the following steps: all samples were derivatised with acetic anhydride and extracted with toluene. The calibration curve was determined with aqueous solutions containing 0.1, 1.0, 6.25, and 25 mg L^{-1} of the three analytes BTri, 4-TTri, and 5-TTri. These three solutions, high-purity water serving as a blank, and the prepared samples were spiked with an internal standard solution (5,6-dimethyl-benzotriazole in sodium carbonate solution) containing the same concentration as used for the calibration curve. The extracts were analyzed by GC tandem mass spectrometry (GC-MS/MS) on a Saturn 2200, Varian

(Agilent Technologies Deutschland GmbH, Böblingen, Germany) equipped with an ion trap. Limit of quantification and limit of detection for this setup were 0.1 ng L^{-1} and 0.01 ng L^{-1}, respectively. Compound separation was accomplished on a VF-5ms column from Varian (30 m x 0.25 mm, film thickness 0.25 μm) perfused by helium as the carrier gas at a constant flow rate of 1.5 mL min^{-1}. The temperature profile started at 65°C (held for 4 min), was increased by 12°C min^{-1} to 200 °C, and was finally set to 300°C at a rate of 40°C min^{-1} (held for 6 min). Operation mode of the MS/MS was resonance excitation of the characteristic precursor ions of the analytes and the internal standard. Injection was performed splitless ranging from 1.0 to 9.0 μL sample volume (large volume liner, Varian 1079 programmable injector). Blanks were analyzed to check for possible contaminations of the experimental samples and to verify the accuracy of the method itself.

2.3 RESULTS AND DISCUSSION

2.3.1 Evaluation of UV-AM

Evaluation of the UV-AM method was performed regarding the following aspects: a) fate of the parent substances by monitoring the change in absorbance due to removal, b) screening for potential transformation products with spectral scans and c) optimization of cultivation media to meet the requirements for application in UV-AM.

2.3.2 Parent Substances

The spectra of the selected compounds were taken in high-purity water to find maximum absorbance and to test whether the used concentrations show sufficient absorbance values for reliable measurements (Figure 2.1 A). Calibration followed to evaluate the compounds behavior in plastic microplates and their absorbance values at different concentrations both in high-purity water (Figure 2.1 B) and the used media (Figure 2.2 for SMX; BTs not shown as their pattern was the same). Absorbance curves and absorbance maxima were different for the xenobiotic compounds analyzed (Table 2.2, Figure 2.1) and an optimal absorbance range for direct measurement was obtained for concentrations from 2-20 mg L^{-1} for all tested xenobiotics.

Table 2.2 – Maximum absorbance (abs_{max}) wavelength of tested xenobiotic compounds and corresponding absorbance value for 10 mg L^{-1}

analyzed compound	absorbance maximum (abs_{max})[nm]	absorbance value abs_{max} for [10 mg L^{-1}]
BTri	253/260	0.52
TTri	263	0.49
4-TTri	262	0.48
5-TTri	264	0.49
SMX	257	0.57

Maximum absorbance values, correlating with the used concentration of xenobiotic compounds, should lay within 0.2 and 1.0 for optimal measurement according to the plate reader setup. Otherwise dilution or higher sample volumes, to increase thickness, have to be used. Absorbance maxima for the used xenobiotics were evaluated (Table 2.2) and subsequently used for single wavelength measurements. Thus, SMX was measured at 257 nm and the BTs at 262 nm (mean value, Figure 2.1). It was not possible to screen for biodegradation in setups containing more than one of the tested xenobiotic compounds as they could not be separately detected in mixed setups (see Figure 2.1 A, BTs).

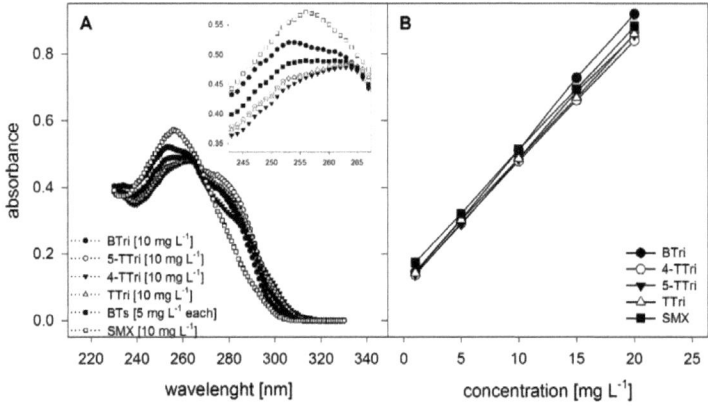

Figure 2.1 A and B – A) UV-absorbance spectra of tested xenobiotic compounds in pure water at 10 mg L^{-1}.
Absorbance maxima are given in Table 2.2. B) Calibration curves of BTri, 4- and 5-TTri, TTri and SMX in pure water. Concentrations were 1.0, 5.0, 10.0, 15.0 and 20.0 mg L^{-1}. Absorbance was measured at 257 nm (SMX) and 262 nm (BTs). Mean values (n=3) are shown, standard deviations too small to be shown (below 1%). Solid lines were fitted by a linear function with R^2 values being >0.99.

Therefore, only one compound per setup could be monitored. Subsequent calibration was performed using 257 and 262 nm to approve the reliability of UV-AM with different concentrations and different media. SMX and the BTs absorbance values behaved linear related to concentration in pure water (Figure 2.1 B).

When tested in different media (Figure 2.2 for SMX, BTs same behavior, data not shown), the media composition showed no effect on absorbance linearity but increased the background absorbance (Figure 2.2) in case of R2A-UV as it contains a high amount of complex nutrients. Substrate consumption and growth of biomass influences on UV-AM could be ruled out by relating xenobiotics containing approaches to blank setups without xenobiotics. MSM and MSM-CN showed the same low absorbance as high-purity water as they just contained mineral salts. Therefore, detection of all tested xenobiotic compounds was possible in pure water as well as in the used media by UV-AM.

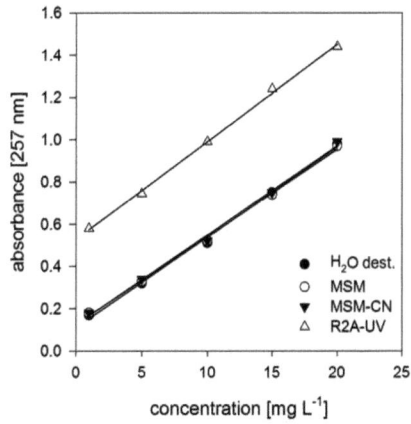

Figure 2.2 - Effect of used media on UV-absorbance.
Shown are, as all other compounds behaved similarly, calibration curves for SMX in pure water and the three media R2A-UV, MSM-CN and MSM (Table 2.1). SMX concentrations were 1.0, 5.0 and 10.0 mg L^{-1}. Mean values (n=3) are shown, standard deviations below 1%.

2.3.3 Screening for Transformation Products

During biodegradation, the molecular structure of the xenobiotic compounds is changed and transformation products might be formed as already shown for SMX [50, 105, 106]. Thus, a change in absorbance might happen that could be observed by spectral scans. Screening for potential transformation products that might exhibit different spectra were performed whenever fast biodegradation was detected, i.e. a significant decrease in absorbance values was observed. Therefore, unnaturally high concentrations up to 100 mg L^{-1} were tested as transformation products normally appear in very low amounts [66]. Unfortunately, even if it should theoretically be possible to detect transformation products by UV-AM spectral scans, it was, under the given experimental conditions, not possible to screen for potential biodegradation products/intermediates. Even for SMX, where already some biotransformation products are known [50, 103, 105], no detection was possible as UV-AM, in this setting, was not sensitive enough to detect transformation.

2.3.4 Optimizing Media for UV-AM

Measurements at xenobiotic-specific wavelengths and under specific laboratory conditions allowed detection of biodegradation in concentrations above 1.0 mg L^{-1} (Figure 2.1 B) as absorbance values are related to compounds concentration. It was possible to detect relative changes in the compounds concentration but optimizations were necessary as the used media strongly influenced UV-AM in two ways: 1) High background. Especially original R2A media showed high background absorption due to the nutrient composition. 2) Biomass growth. Available nutrients led to biomass growth and thus increased absorbance (Table 2.1) as at 257 and 262 nm also DNA, cells, and cell-particles show an absorbance [107]. Media containing components, e.g. yeast extract or peptone, show high media background absorbance and foster biomass growth but are not well suited for UV-AM as background absorbance can superimpose potential biodegradation. Nevertheless, these media enable growth of a high diversity of organisms and contain complex nutrient sources that may enhance cellular activity and thus biodegradation. As already shown for SMX, sufficient nutrient availability can foster biodegradation [50].

Therefore, original R2A media, being necessarily used for its complex nutrient composition had to be optimized as it showed absorbance values around 3.40 units. The two MSM media already showed very low absorbance values around 0.08 and could be used without changes. The exchange of nutrients in original R2A without changing the total nutrient concentration led to development of R2A-UV media and a decrease in background absorbance from 3.40 to 0.54 units (composition see Table 2.1, exchanged nutrients in bold). In media or water without biomass, xenobiotics' concentrations could be measured, but as growth of organisms influences absorbance exact concentrations' determination could not be provided with extensive biomass growth in the setup. Therefore, it was necessary to test the influence of biomass on UV-AM. Nevertheless UV-AM was applicable for fast and inexpensive detection of relative changes in xenobiotics´ concentrations within the applied media given the restriction that only one compound can be used per biodegradation setup.

2.3.5   Screening ASCs for Xenobiotic Biodegradation

2.3.5.1   UV-AM Evaluation with Biomass

After detection of a decrease in UV-absorbance indicating biodegradation, these setups were analyzed by LC-UV for SMX and GC-MS/MS for BTs to validate UV-AM in different matrixes and relate absorbance with exact concentrations. If a screening in biomass-containing setups is possible, UV-AM can significantly increase performance while reducing analytical costs (Table 2.3).

Table 2.3 – Comparison of UV-AM with LC-UV and GC-MS/MS measurements regarding time needed for analyzes and costs per sample

	benzotriazoles			sulfamethoxazole		
	UV-AM	LC-UV	GC-MS/MS	UV-AM	LC-UV	GC-MS/MS
time per sample [min]	<1	-	20	<1	5	-
costs per sample [€]	0.12	-	40-60	0.12	30-40	-
measurement of biodegradation	yes	-	yes	yes	yes	-
measurement of concentrations	difficult	-	yes	difficult	yes	-
labor intensive	no	-	yes	no	yes	-
time intensive	no	-	yes	no	yes	-
monitoring biodegradation	yes	-	yes	yes	yes	-

All ASCs were able to biodegrade most of the tested compounds under aerobic conditions in R2A-UV media, regardless the initial MLSS concentration. Degradation was observed with xenobiotic concentrations of 10 mg L^{-1} and it was possible to compare different initial biomass concentrations. Incubation times until degradation was detectable with UV-AM varied with the tested compound and initial biomass concentration. Chosen setups were analyzed in a biodegradation time series by both UV-AM and LC-UV or GC-MS/MS to test the reliability of UV-AM results (Figure 2.3).

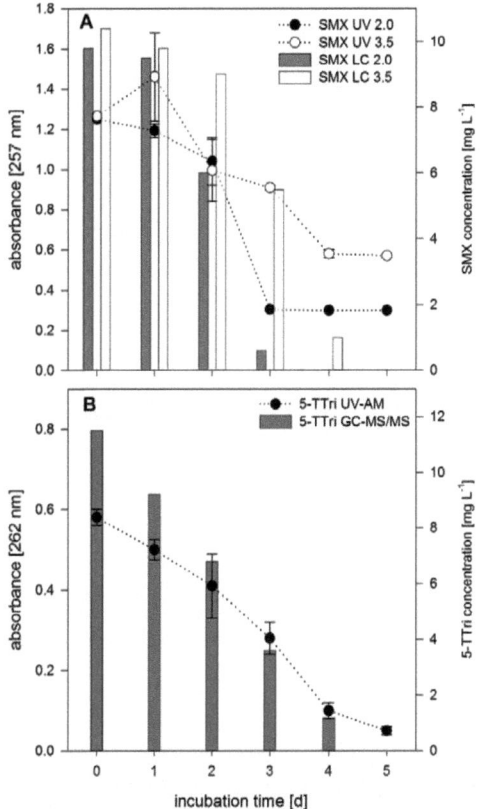

Figure 2.3 - Correlation of UV-AM results with LC-UV and GC-MS/MS analyses. A) Shown is SMX biodegradation in R2A-UV media with two different MLSS concentrations (2.0 and 3.5 g L^{-1} MLSS). Initial SMX concentration 10 mg L^{-1}. Dotted lines show the measured SMX biodegradation detected with UV-AM, columns represent SMX concentration measured with LC-UV. B) The same is shown for 5-TTri (initial concentration 10 mg L^{-1}, measured with GC-MS/MS) biodegradation with initial MLSS concentration of 0.5 g L^{-1}. Shown are mean values of duplicate experiments with error bars indicating standard deviations (n=2).

Double measurements using UV-AM and LC/GC methods showed that SMX as well as BT biodegradation could be detected by UV-AM as the decrease in absorbance values could be correlated to the declined SMX and BTs' concentrations determined by LC and

GC measurements. Even in experiments with a high biomass concentration of 3.5 g L^{-1} MLSS (Figure 2.3 A), the decrease in UV-AM fitted the measured values in LC-UV, but the background was increased significantly probably due to solved compounds originating from the AS inoculum (see Figure 2.3 A, at day five background absorbance of 0.56 units with all SMX being biodegraded). In contrast, the 5-TTri setup carried out with only 0.5 g L^{-1} MLSS (Figure 2.3 B) after 5 days of incubation showed hardly any background absorption. Therefore, this lower MLSS concentration seems to be more useful for UV-AM. Nevertheless, high MLSS concentrations, as they are used in WWTPs, had to be tested to evaluate if UV-AM can also be used directly for screening ASC for their biodegradation potential. Even with high biomass density, a proper sedimentation provided, UV-AM represented a fast and cost-effective way to screen for biodegradation (Table 2.3). Screening 96 samples with UV-AM, including sample preparation, took around 30-40 min and required costs of around 12 €. Compared to LC-UV or GC-MS/MS this is a significant saving in time and money when concentrations are high enough in laboratory experiments (Table 2.3).

2.3.5.2 UV-AM for Detection of Xenobiotics' Biodegradation under Different Nutrient Conditions

Another advantage of UV-AM was the possibility to screen for differences regarding nutrient concentration and biomass concentration. Figure 2.4 A and B show for SMX (BTri not shown as it behaved similarly) two setups differing in nutrient concentration (MSM or MSM-CN media) and biomass amount (2.0

or 3.5 g L^{-1} MLSS). It became clear that in MSM-CN biodegradation did not require any acclimation, it started right after inoculation, presumably due to the microorganisms´ higher activity and showed a more constant and linear removal. In MSM an acclimation period of almost four days was observed followed by a very rapid SMX removal within four days.

Another finding was that a higher biomass concentration positively affected biodegradation as setups with 3.5 g L^{-1} MLSS worked slightly better in both duplicates. This effect was rather clear for MSM-CN but with nutrients shortage in MSM, biomass concentration did not have that effect on biodegradation as the general activity was lower and thus more time for acclimation required.

In both setups SMX was biodegraded in around 8 days. Additional experiments were performed regarding nutrient composition and initial biomass concentration (Table 2.4). Only initial xenobiotics´ concentrations were measured by LC/GC and a large number of setups were screened for their biodegradation potential. As Table 2.3 also shows, a lot of setups could not be measured as background absorption was too high for detection of changes in absorbance values. Especially setups with 1 mg L^{-1} initial xenobiotics´ concentration were superimposed by higher background absorbance due to biomass concentration.

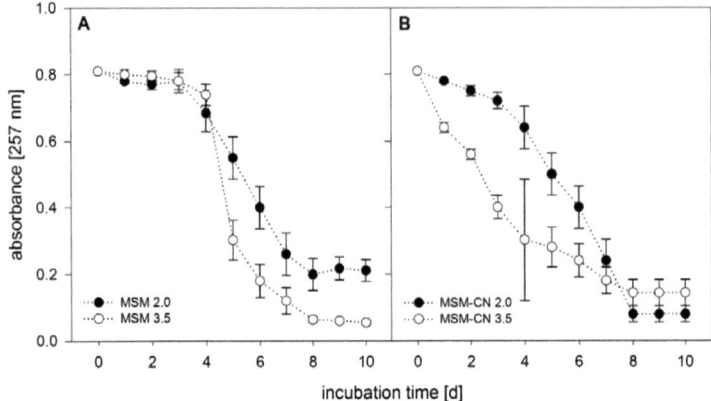

Figure 2.4 A and B - Time series of SMX biodegradation with activated sludge communities in MSM-CN (A) and MSM (B) as detected by UV-AM.
Initial SMX concentration 20 mg L^{-1}. Shown are mean values of SMX absorbance in duplicate experiments with error bars indicating standard deviations (n=2).

Table 2.4 - Biodegradation and time needed for absorbance decrease in the different ASC setups using different media, biomass concentration and initial xenobiotics´ concentrations. Shown are BTri, 5-TTri and SMX. 4-TTri was not biodegraded and is thus not shown.

Xenobiotic	concentration [mg L^{-1}]	initial MLSS [g L^{-1}]	decrease of absorbance values[1]			lowest absorbance value achieved after [d][2]		
			R2A-UV	MSM-CN	MSM	R2A-UV	MSM-CN	MSM
BTri	1	0.5	Byes	yes	yes	10	12	13
	1	2.0	n.d.	n.d.	n.d.	n.a.	n.a.	n.a.
	1	3.5	n.d.	n.d.	n.d.	n.a.	n.a.	n.a.
	10	0.5	yes	yes	yes	15	16	18
	10	2.0	yes	yes	yes	14	12	13
	10	3.5	n.d.	n.d.	n.d.	n.a.	n.a.	n.a.
5-TTri	1	0.5	yes	yes	yes	10	8	7
	1	2.0	n.d.	n.d.	n.d.	n.a.	n.a.	n.a.
	1	3.5	n.d.	n.d.	n.d.	n.a.	n.a.	n.a.
	10	0.5	yes	yes	yes	10	9	9
	10	2.0	yes	yes	yes	8	8	8
	10	3.5	n.d.	n.d.	n.d.	n.a.	n.a.	n.a.
SMX	1	0.5	yes	yes	yes	4	3	4
	1	2.0	n.d.	n.d.	n.d.	n.a.	n.a.	n.a.
	1	3.5	n.d.	n.d.	n.d.	n.a.	n.a.	n.a.
	10	0.5	yes	yes	yes	6	5	8
	10	2.0	yes	yes	yes	3	8	8
	10	3.5	yes	yes	yes	4	8	8

n.d. – no biodegradation determination possible, background due to biomass concentration too high
n.a. – not applicable as no biodegradation could be detected by UV-AM
[1] change in UV-AM indicates biodegradation
[2] time needed to achieve a stable value in UV-AM after a change in absorbance

2.4 CONCLUSIONS

A UV absorbance measurement technique was developed as a pre-test for the evaluation of xenobiotics' biodegradation potential. UV-AM approved time- and cost-saving, reproducible and reliable to screen a large number of different laboratory setups within short time. Obtaining fast results in biodegradation experiments is possible with the premise that the concentrations of the tested compounds are within the range of 1.0 to 20.0 mg L^{-1}. Additionally, specific media were used to reduce both, background absorption and biomass growth. As a limitation only one compound could be used per biodegradation setup, e.g. benzotriazole, 4-methyl- and 5-methyl-benzotriazole or sulfamethoxazole could be analyzed by UV-AM to identify biodegradation potential. It was possible to detect xenobiotic biodegradation in reactors inoculated with activated sludge. Validation of UV-AM results was performed by comparing the values with either LC-UV for SMX or GC-MS/MS for benzotriazoles. Due to the use of microplates an "easy to handle" system allowing high throughput screening was established. 96 well or even 384 well plate formats can be used, requiring less time and saving laboratory costs. Most important, it is possible to pre-select biodegrading communities, optimize the conditions for biodegradation with respect to nutrient concentration and activated sludge inocula and monitor biodegradation over time and to screen pure culture setups for their biodegradation potential. This minimizes required time and costs for LC or GC measurements as only setups showing a decrease in absorbance need to be further analyzed and can be used in subsequent biodegradation experiments. However, high biomass concentrations turned out disadvantageous as they created a high background absorbance and thus superimposed UV-AM. Therefore, UV-AM should not be used for exact concentrations measurements. Nevertheless, relative changes in xenobiotics' concentrations could be detected which is usually sufficient as for pre-selections or monitoring approaches exact concentrations are not needed.

Chapter 3

Monitoring Benzotriazoles: a One Year Study on Concentrations and Removal Efficiencies in Three Different Wastewater Treatment Plants and their

Benzotriazole (BTri), 4- and 5-methyl-benzotriazole (4-TTri, 5-TTri) were monitored over one year in three wastewater treatment plants (WWTP): a membrane bioreactor (MBR-MH) and two conventional activated sludge systems (CAS-E, CAS-M). Influent/effluent concentrations, treatment stages removal efficiencies and impact on the receiving rivers were monitored. 5-TTri was removed best with a mean of 75% in the WWTPs mainly by biodegradation in the aeration tanks compared to BTri (mean total removal of 45%) and 4-TTri (mean total removal of 15%) that showed significant lower eliminations. High removal fluctuations for all three benzotriazoles occurred over the four seasons with lowest removal during winter. All three WWTPs constituted a point source for BTs in the aquatic environment. Mean downstream benzotriazole concentrations were 3.58 µg L^{-1} for MBR-MH and 1.1 µg L^{-1} for CAS-M. CAS-E only slightly increased downstream concentrations to 0.64 µg L^{-1} as the receiving river was already contaminated with benzotriazoles (mean 0.49 µg L^{-1}) from hydropower. BTri and 4-TTri have the potential to accumulate in the aquatic systems.

3.1 INTRODUCTION

The xenobiotic compounds benzotriazole (BTri) as well as 4- and 5-tolyltriazole (4-TTri, 5-TTri), summarized here as BTs, are polar micropollutants commonly used as corrosion inhibitors [60]. Commercially available tolyltriazole (TTri), a mixture consisting of 4- and 5-TTri (40/60%) is widely used in metal finishing and corrosion protection in cooling systems [108]. BTri is found in dishwashing detergents [83] but also as additive to aircraft deicing or breaking fluids[88], for silver protection [63] and UV filter and/or sun screen agent [109].

Production volumes of 9,000 tons per year in the United States categorizes these BTs as 'high production volume chemicals' [99].

Their widespread usage, polar nature (logD 1.44 and 1.71 for BTri and TTri, respectively) and their good water solubility of 19.0 (BTri), 6.0 (5-TTri) and 0.25 (4-TTri) g L^{-1} implicates their almost omnipresent occurrence in all larger freshwater aquatic systems across Europe [94]. All three BTs were recently detected in groundwater with concentrations up to 1548 ng L^{-1} [17], river systems with up to 27,393 ng L^{-1} [53] and even the North Sea with 40 ng L^{-1} [110]. Even higher concentrations of 128 and 198 mg L^{-1} for BTri and TTri, respectively, were found in a water monitoring well at an international airport [111].

Concentrations above 0.97 mg L^{-1} for BTri and 0.40 mg L^{-1} for 5-TTri already showed adverse effects in a chronic study with the aquatic organism *Daphnia galeata* [112]. Concentrations above 40 mg L^{-1} (BTri) and 6 mg L^{-1} (5-TTri) showed toxic effects in Microtox®

tests (*Vibrio fischeri* light emission as toxicity test system, Azur Environmental, CA) [113]. Therefore, BTs are regarded as potentially hazardous for the aquatic system [114]. Optimizing these compounds removal during wastewater treatment to reduce their discharge into aquatic systems is thus implicitly needed. Wastewater treatment plants (WWTP) approved incapable to completely remove these compounds [101, 102] making them one major point source into the aquatic environment [63].

The present study is screening the removal efficiency of the micropollutants BTri, 5-TTri and 4-TTri in three WWTPs with membrane (MBR-MH) and conventional activated sludge treatment (CAS-E, CAS-M) over one year. BTs concentrations were monitored in WWTPs influent, before and after different stages regarding the contribution of the treatment stage for their elimination. This allows evaluating the potential elimination process that occurs in the treatment stage. To show WWTPs impact of the aquatic system, their up- and downstream effluent concentrations were compared in the receiving rivers.

3.2 MATERIALS AND METHODS

3.2.1 Wastewater Treatment Plants and Receiving Water Bodies

Three WWTPs were chosen for this study being different in their treatment techniques (Table 3.1 and Figure 3.1). A membrane bioreactor (MBR-MH), a conventional activated sludge plant with intermittent nitrification/denitrification (CAS-E) and a two stage (stage one: high load stage for C-elimination and pre-denitrification;

stage two: low load stage for nitrification) activated sludge plant (CAS-M).

Table 3.1 – Characteristics of wastewater treatment plants (WWTPs) analyzed and their receiving rivers

WWTP	type of treatment	HRT [h]	MLSS [g L^{-1}]	sludge age [d]	F/M ratio sludge load [BOD$_5$ kg^{-1} MLSS d^{-1})]	fraction of industrial wastewater [%]	receiving river system
MBR-MH	membrane filtration, pore size 0.1 µm	10-14	8-12	30-40	<0.15	0-2	River Gailach
CAS-M stage 1	2-stage activated sludge treatment	1-3	3-4	1-2	<0.64	45-55	River Isar
CAS-M stage 2		2-5	3-4	8-10	<0.05		
CAS-E	intermittent nitrification/ de-nitrification	3-6	3-4	1-2	<0.30	20-30	Middle Isar-Channel

PE – population equivalents
HRT – hydraulic retention time
MLSS – mixed liquor suspended solids
F/M ratio – food to microorganism ratio

All downstream sampling points for the receiving rivers were calculated according the formula for mixing length $L_m = 8,52*U*B^2/D$ with U: flow velocity [m s^{-1}], B: river width [m], D: river depth [m] to ensure a homogenous mixing of river water with WWTP effluent [115]. All WWTP samples (Figure 3.1) were collected by 24 h composite automatic samplers. Grab samples were collected from the receiving rivers. All samples were cooled and frozen at -20°C before analysis.

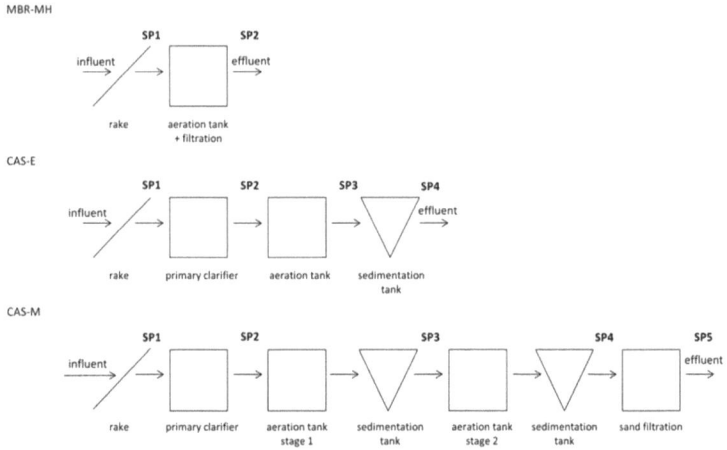

Figure 3.1 – Flow charts of the monitored WWTPs MBR-MH, CAS-E and CAS-M including the different treatment stages and sampling points (SP).
MBR-MH, a membrane bioreactor, was sampled before and after the biological/filtration stage. CAS-E, a conventional activated sludge plant operating with intermittent nitrification/denitrification, was sampled at four points. CAS-M was sampled at five points and operates a two stage biological treatment with stage 1 being the high load stage for carbon elimination and pre-denitrification, and stage 2 the low load stage for nitrification.

MBR-MH, equipped with membrane filtration, discharges into the river Gailach, a very small river with a mean volumetric flow rate of around 20 L s^{-1} over the year. In dry summer seasons it receives almost all its water from the WWTP. Approx. 6 km downstream the WWTP, it drains away into a Karst zone to reappear after approx. 7 km to form the lower Gailach spring. Therefore, the discharge of pollutants or microorganisms from the WWTP has to be reduced efficiently as the groundwater, supplied by the Gailach, is used as drinking water. MBR-MH was thus equipped a with membrane filtration. Samples were collected from the WWTP influent (SP1), immediately after the rake and the effluent

(= permeate, SP2). Additionally, Gailach water samples were collected upstream (48.839316 N, 10.870032 E) and 500 m downstream (48.839408 N, 10.87221 E).

CAS-E, operated with conventional activated sludge treatment and equipped with a sedimentation tank, discharges into the Middle Isar-Channel, an artificial channel operating hydroelectric power plants. River water was sampled upstream the WWTP (48.36655 N, 11.88581 E) and after 2 km downstream (48.384922 N, 11.91597 E). Samples of the WWTP itself were collected after rake (SP1), primary clarifier (SP2), aeration tank (SP3) and sedimentation tank (SP4).

CAS-M, a conventional two-stage activated sludge treatment plant discharges right into the river Isar. This river, with a mean flow rate of 175 $m^3 s^{-1}$ and a total length of 295 km, serves during the summer season as recreation and bathing area [116]. Water samples were collected before WWTP discharger (48.29479 N, 11.698534 E) and downstream (48.313149 N, 11.699789 E). WWTP samples were collected after rake (SP1), primary clarifier (SP2), aeration tank stage 1 and sedimentation tank (SP3), aeration tank stage 2 and sedimentation tank (SP4) and the effluent (SP5).

3.2.2 Chemicals

1-H-benzotriazole (BTri; 99,8 % purity; CAS 95-14-7), 4-methyl-benzotriazole (4-TTri; CAS 29878-31-7) and 5-methyl-benzotriazole (5-TTri; CAS 136-85-6) were provided by Cimachem GmbH (Kirchheimbolanden, Germany) and served as calibration standards. The surrogate standard 5,6-dimethyl-benzotriazole

(99% purity), sodium carbonate solution, toluol and all other chemicals were purchased from Sigma-Aldrich (Steinheim, Germany).

3.2.3 Sample Preparation and Chemical Analyses of BTs

All WWTP and river water samples were centrifuged (10 min, 8,000 g, 7°C) to remove (waste)-water residues, filled into cleaned and sterilized glass bottles and stored in the dark at -20°C before analysis.

Chemical preparation and determination of the BTs' concentration in the liquid samples were performed according to Liu et al. (2011) with the following steps: all samples were derivatised with acetic anhydride, and extracted with toluene. The calibration curve was determined with aqueous solutions containing 0.1, 1.0, 6.3 and 25.0 µg L^{-1} of BTri, 4-TTri, and 5-TTri. These three solutions, high-purity water serving as a blank, and the prepared samples were spiked with an internal standard solution (5,6-dimethyl-benzotriazole in sodium carbonate solution) containing the same concentration as used for the calibration curve. The extracts were analyzed by gas chromatography followed by tandem mass spectrometry (GC/MS-MS) on a Saturn 2200 by Varian (Agilent Technologies Deutschland GmbH, Böblingen, Germany) equipped with an ion trap. Limit of quantification for this setup was 0.01 µg L^{-1}.

3.3 RESULTS AND DISCUSSION

3.3.1 Influent and Effluent Concentrations and Total Removal Efficiencies of the WWTPs

Benzotriazoles wastewater influent and effluent concentrations were significantly different for the three WWTPs and showed a fluctuation over one year (Figure 3.2, Table 3.2).

Table 3.2 – Mean influent and effluent concentrations with standard deviations of BTri, 5-TTri and 4-TTri over one year determined for the three WWTPs MBR-MH, CAS-E and CAS-M.

WW-TP	total BTs [µg L^{-1}]		BTri [µg L^{-1}]		5-TTri [µg L^{-1}]		4-TTri [µg L^{-1}]	
	in-fluent	ef-fluent	in-fluent	ef-fluent	in-fluent	ef-fluent	in-fluent	ef-fluent
MBR-MH	11.65± 10.12	7.09± 4.63	5.89± 4.01	3.47± 1.60	2.22± 2.21	0.41± 0.30	3.54± 3.80	3.20± 2.70
CAS-E	35.83± 26.54	12.78± 13.70	25.56± 20.70	8.24± 10.70	5.41± 3.00	0.39± 0.25	4.86± 2.78	4.16± 2.71
CAS-M	18.09± 10.67	13.37± 5.93	11.46± 5.22	9.26± 3.34	3.05± 2.89	0.73± 0.37	3.59± 2.55	3.38± 2.21

Highest BTs mean wastewater concentrations of 35.83 µg L^{-1} and a peak concentration of 101.90 µg L^{-1} were found in CAS-E that receives water from the Munich airport. Lower concentrations with a mean of 18.10 µg L^{-1} and a peak concentration of 36.90 µg L^{-1} were detected in CAS-M and the lowest in MBR-MH, a small treatment plant with hardly any industrial influence, with a mean of 11.65 µg L^{-1} and a peak concentration of 35.70 µg L^{-1}. Therefore, all BTs in the wastewater from MBR-MH likely originate from their use in dishwashing formulations as other studies already found that wastewater discharge from households contains significant concentrations of BTs [63, 83].

CAS-M receives almost two times higher concentrations of BTri (11.46 µg L^{-1}) compared to MBR-MH (5.89 µg L^{-1}) but showed similar values as MBR-MH for both, 5-TTri and 4-TTri (around 3 µg L^{-1}). Therefore, industry, contributing up to 50% of the total influent volume of CAS-M, might mainly use BTri for corrosion protection instead of the other two compounds.

In CAS-E, the concentrations for 5-TTri and 4-TTri were similar low as detected in MBR-MH and CAS-M whereas the concentration of BTri was more than two times higher compared to CAS-M and almost five-times higher compared to MBR-MH. CAS-E receives up to 30% industrial wastewater solely produced by the Munich airport. The airport wastewater contains very high BTri concentrations resulting in high inflow concentrations of around 25.56 µg L^{-1} BTri and distinctly lower concentrations of 5.41 µg L^{-1} for 5-TTri and 4.86 µg L^{-1} for 4-TTri. It is known from literature that airport wastewaters contain high amounts of BTs, especially in the winter seasons when aircraft-deicing is applied [87, 88, 117]. As the Munich Airport is not using BTs any more as additive to aircraft-deicing formulations, a strong seasonal dependence was not found in the present study and BTs influent concentrations were quite stable over the year. One exception was October with a high BTri influent concentration of 91 µg L^{-1} that could be correlated to the high airport wastewater concentration of 101 µg L^{-1} BTri. Therefore, other input sources for BTri must exist that contribute to the higher BTri concentration compared to MBR-MH. Generally, BTri was always the major influent component of the BTs whereas 5-TTri and 4-TTri were found in similarly lower concentrations.

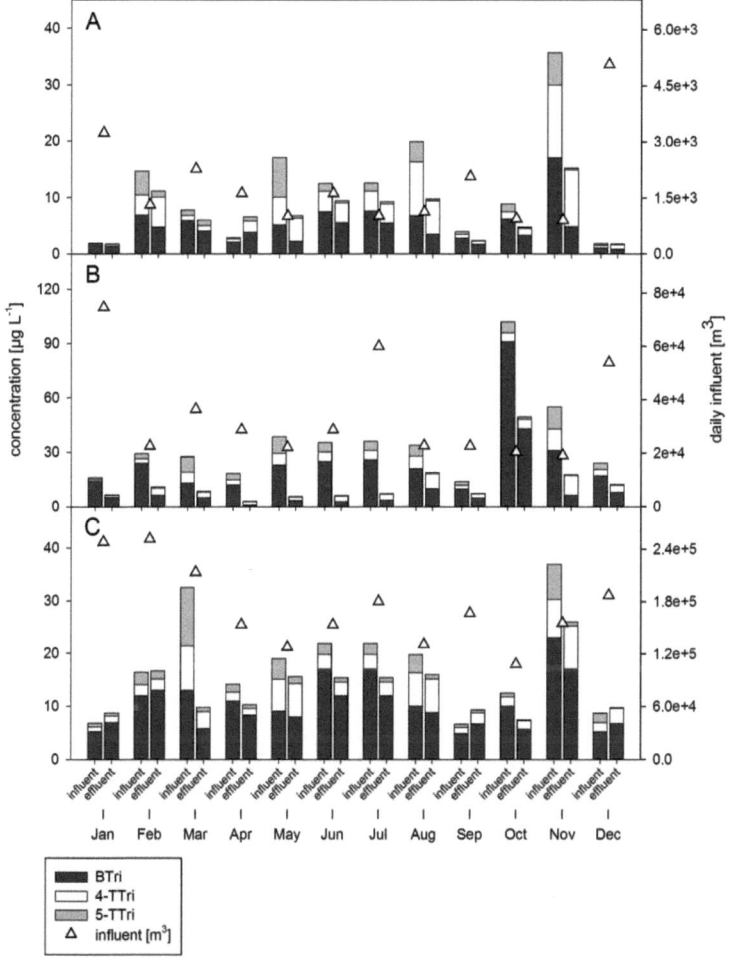

Figure 3.2 – Influent and effluent concentrations of the three BTs BTri, 4-TTri, and 5-TTri and the daily influent volume of the three WWTPs: A) MBR-MH, B) CAS-E, and C) CAS-M.

All three WWTPs were able to remove these BTs with varying efficiencies, as effluent concentrations were mostly lower than influent concentrations. A similar removal pattern as already reported by other studies with best removal for 5-TTri with up to 69% followed by BTri with up to 58% and worst for 4-TTri with up to 34% [85, 101], was also found here.

The mean BTs removal efficiencies found in this study, regarding the three WWTPs, showed a similar pattern with a good removal of 5-TTri (mean 75%) followed by a weaker removal of BTri (mean 45%) and a low removal of 4-TTri (mean 15%) observed here (Figure 3.3).

Strong fluctuations in the analyzed WWTPs were observed for 4-TTri and BTri that ranged from almost complete to no removal at all. Best elimination for all three BTs was observed in CAS-E that, especially for 5-TTri and BTri, showed a high and consistent removal. Highest fluctuation in the removal efficiency was found for MBR-MH that might be attributed to the small size of the plant.

Furthermore, the daily influent volume (right axis, Figure 3.2) significantly influenced removal efficiency. In January and December, WWTPs received higher influent volumes due to rainfall resulting in significantly reduced BTs influent concentrations and a decreased removal in all three WWTPs. This low removal might be attributed, to some extent, to lower temperatures reducing microbial activity. However, hydraulic retention time (HRT) was more important for removal as observed for February and November: Influent volume was significantly lower, compared to January and December, resulting in good removal of the BTs as the organisms likely had more time to remove these compounds even at low temperatures.

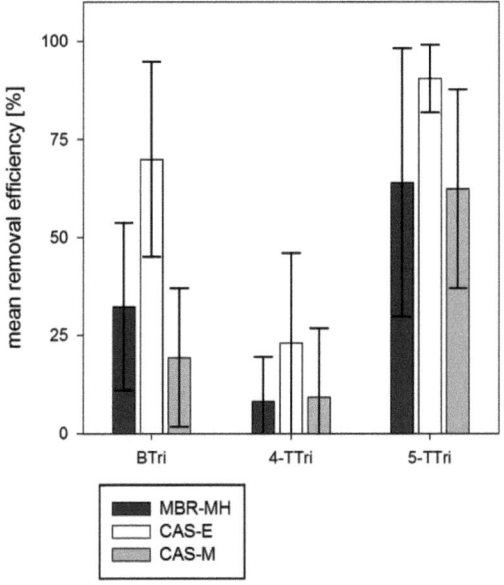

Figure 3.3 – Mean BTri, 4-TTri and 5-TTri removal efficiencies with standard deviation (n=12) over one year of the three WWTPs, MBR-MH, CAS-E and CAS-M.

3.3.2 Removal Efficiencies of the Wastewater Treatment Stages

BTs removal within the different treatment stages (Figure 3.1) were analyzed (Figure 3.4) and showed that 5-TTri is best removed followed by BTri and 4-TTri independent of treatment stage and WWTP system.

Sorption onto biomass has to be neglected for all three BTs during wastewater treatment [26]. Main removal processes are biodegradation in the aeration stages and photolysis, which might be the dominant elimination process in sedimentation, as the less turbid conditions allow photodegradation. Nevertheless, for CAS-E, an impact of the sedimentation tank can be seen, but a clear separation of biodegradation in the aeration tank and an abiotic (photolytic) removal in the sedimentation is not possible. The same is valid for the primary clarifier which contributes but not with a clear pattern.

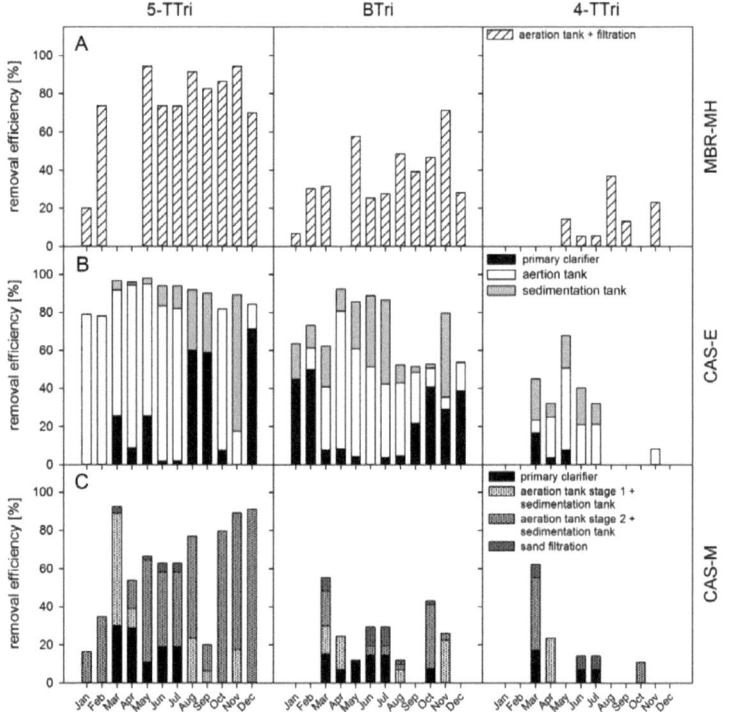

Figure 3.4 – Removal efficiencies per sampled treatment stage of 5-TTri, BTri and 4-TTri over one year analyzed for the WWTPs A) MBR-MH, B) CAS-E and C) CAS-M.

For MBR-MH, the impact of aeration tank and filtration stage on BTs removal (Figure 3.4 A) could not be evaluated but a laboratory filtration experiment with the membrane material used in MBR-MH showed that the membrane itself had no influence on BTs removal. Therefore, BTs removal in MBR-MH in the aeration tank together with filtration unit (Figure 3.4 A) is mainly attributed to biodegradation. Good annual 5-TTri elimination was observed, except in March and April, where this plant had general problems due to membrane cleaning and thus showed no removal at all. A similar pattern but with a generally lower removal was found for BTri while 4-TTri showed very weak elimination. A removal of around 40% was observed in August only. In the summer season, the WWTPs' HRTs were higher as the influent volume was lower whereby 4-TTri might be better removed. MBR-MH lacks a primary clarification and sedimentation thus limiting removal to biodegradation. Furthermore, the total retention time of this small plant was shorter compared to the other two WWTPs giving the organisms less time for biodegradation. This explains why BTri was only poorly removed although biological removal was shown in laboratory studies [90] and could even be completely removed after sufficient incubation time (see also Chapter 4).

In CAS-E, three treatment stages were sampled separately and their contribution to total removal efficiency evaluated (Figure 3.4 B). 5-TTri is mainly removed in the aeration tank but to a lesser amount also in the primary clarifier and sedimentation probably due to biodegradation. In August and September, no 5-TTri was removed in the aeration tank but a high removal efficiency of around 90% was achieved in primary clarifier and sedimentation. Laboratory biodegradation experiments showed that 5-TTri is easily biodegraded [90] and retention in the primary clarifier might be enough to biodegrade 5-TTri. CAS-E also showed around 70% removal for BTri with higher impact of the primary clarifier compared to CAS-M.

Finally, CAS-M showed the lowest BTs removal efficiencies of the three WWTPs (Figure 3.4 C). This plant operates two aeration stages, a high load stage 1 (carbon removal, pre-denitrification) and a low load stage 2 (nitrification) that are different in both food to microorganism ratio and sludge age (Table 3.1). Removal of 5-TTri occurred mainly in stage 2 (lower carbon concentrations). That indicates stage 2 to be better suited to remove persistent organic compounds as A) the food to microorganism ratio is lower forcing the organisms to use not readily degradable compounds being available only in $\mu g\ L^{-1}$ concentrations. B) A higher sludge age favors higher sludge diversity and slow growing organisms supporting BTs degradation. C) A higher HRT gives the organisms more time to get adapted to utilize poorly degradable xenobiotic compounds like BTs [82]. In contrast, BTri removal was low in all treatment stages and no removal was observed during winter. It is concluded that the main removal process for BTri might be biodegradation explaining why lower temperatures, i.e. reduced microorganisms' activity as typical for the winter season, reduced removal efficiency.

3.3.3 WWTPs Discharge into Receiving Rivers

BTs concentrations were measured in the receiving rivers to evaluate their impact on aquatic systems (Figure 3.5, Table 3.3). MBR-MH (Figure 3.5 A) and CAS-M (Figure 3.5 C) constitute one major point source for BTs into the aquatic environment as the detected BTs concentrations increased significantly downstream the WWTPs. CAS-E (Figure 3.5 B) showed no such effect as the Middle Isar-Channel was, in some months, higher contaminated with BTs (peak concentration of 1.06 µg L^{-1}) than CAS-E effluents. MBR-MH showed the highest impact on the receiving water with BTs concentrations up to 10 µg L^{-1} and a strong fluctuation over the year. The receiving river is very small and the WWTPs effluent contributes, especially during dry summers, to almost 50% of the total river discharge. This effect was observed for the dry months June to November where the river nearly dried up explaining why almost the complete effluent concentration of MBR-MH was found in the aquatic system (Figure 3.5 A compared to Figure 3.4 A). It is shown that MBR-MH effluents significantly increased the amount of BTs in the Gailach River (Table 3.3) especially of BTri and 4-TTri, the two biologically very stable compounds. These compounds might accumulate in the aquatic ecosystem and consequently represent a potential threat for the ecosystem and drinking water supply downstream. In contrast, 5-TTri was well removed by MBR-MH and was readily biodegraded resulting in a low risk for that compound to accumulate and remain in the aquatic system.

Table 3.3 – Mean BTs concentrations with standard deviations measured over one year in the receiving rivers upstream and downstream the corresponding WWTPs.

receiving river system/ WWTP	total BTs [µg L^{-1}]		BTri [µg L^{-1}]		5-TTri [µg L^{-1}]		4-TTri [µg L^{-1}]	
	up-stream	down-stream	up-stream	down-stream	up-stream	down-stream	up-stream	down-stream
Gailach/ MBR-MH	0.36± 0.70	3.57± 3.11	0.19± 0.36	1.66± 1.20	0.01± 0.01	0.16± 0.10	0.16± 0.32	1.76± 1.74
Middle Isar-Channel/ CAS-E	0.50± 0.30	0.64± 0.31	0.35± 0.19	0.43± 0.19	0.02± 0.01	0.03± 0.01	0.12± 0.08	0.18± 0.10
Isar/ CAS-M	0.20± 0.10	1.10± 0.50	0.12± 0.04	0.77± 0.30	0.02± 0.01	0.05± 0.02	0.06± 0.03	0.28± 0.16

A similar impact pattern but with lower concentrations was found for CAS-M that directly discharges into the river Isar (Figure 3.5 C, Table 3.3). This river, although around 10,000 times larger than the Gailach, exhibits similar seasonal fluctuations with low flow volumes during summer and high flow volumes in spring and winter due to rainfall and snowmelt. Again, 5-TTri showed the lowest concentrations around 0.2 µg L^{-1} as it was well removed by the WWTP and might not accumulate due to the river's self-purification potential. BTri was found in highest concentrations (up to 1.4 µg L^{-1}) followed by 4-TTri (up to 0.6 µg L^{-1}) as they were only partly removed in the WWTP.

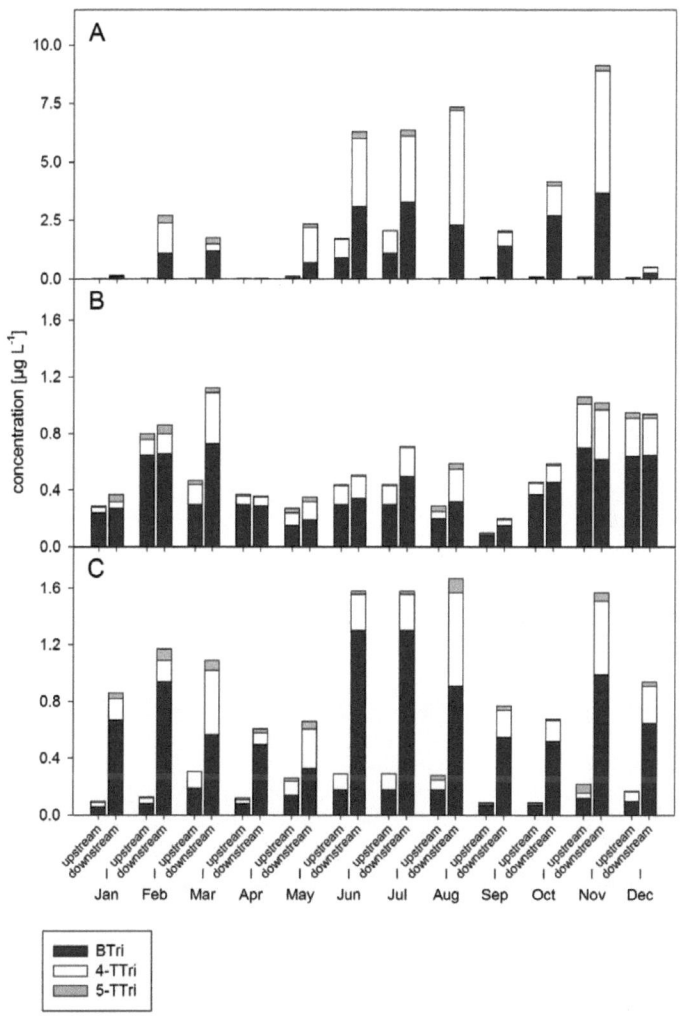

Figure 3.5 – Concentrations of BTri, 4-TTri and 5-TTri upstream and downstream the WWTPs over one year.
BTs influence receiving waters for A) MBR-MH/Gailach, B) CAS-E/Middle Isar-Channel and C) CAS-M/Isar.

CAS-E showed a different impact pattern as the receiving Middle Isar-Channel contained relatively high BTs concentrations up to 1.0

µg L^{-1} upstream the WWTP (Figure 3.5 B, Table 3.3) arising from diffuse entry paths from hydropower plants. This artificial channel mainly constructed for hydroelectricity shows no seasonal fluctuations as its flow volumes are controlled.

The ratio of river flow volume to effluent amount of WWTP CAS-E was around 1:500. Therefore CAS-E's impact on the receiving water was far less as A) the effluent was highly diluted and B) removal efficiency of CAS-E was best for the BTs compared to the other two WWTPs. BTs effluent concentrations were sometimes even lower than these upstream therefore reducing BTs concentrations downstream by dilution only (April, November and December). In the channel upstream and downstream the WWTP, BTri (up to 0.7 µg L^{-1}) and 4-TTri (up to 0.5 µg L^{-1}) were again the dominant species while 5-TTri was found up to 0.04 µg L^{-1}.

However, the mean concentrations over one year downstream the WWTPs increased for all three BTs (Table 3.3) in the three WWTPs. The WWTPs were not able to completely removing these BTs from the aquatic systems whereas further removal processes, e.g. self-purification of the receiving rivers, mixing/dilution effects and photodegradation might be important to finally preventing BTs accumulation in the aquatic systems.

3.4 CONCLUSIONS

In this study, three different wastewater treatment plants (WWTPs) were monitored for their benzotriazole, 5-tolyltriazole and 4-tolyltriazole concentrations in the influent and effluent over one year. Mean influent concentrations were highest for BTri ranging from 5.89 to 25.56 µg L^{-1} followed by 5-TTri and 4-TTri showing similar concentrations from 2.22 to 5.41 µg L^{-1}. Seasonal fluctuations were low regarding influent concentrations but high regarding annual removal efficiency and impact on receiving rivers. 5-TTri was removed preferentially biologically with a mean of 75% while BTri (mean 45%) and 4-TTri (mean 15%) showed less removal. All three WWTPs were a point source for these BTs into the aquatic environment but their impact strongly depended on the upstream water quality within the receiving river system. However, as strong fluctuations in the WWTPs' removal efficiencies occurred, further research on optimizing the removal of all three BTs is implicitly required.

Chapter 4

Xenobiotic Benzotriazoles - Biodegradation under Meso- and Oligotrophic Conditions as well as Different Redox Potentials

Evaluation of activated sludge communities (ASC) biodegradation potential towards benzotriazole (BTri), 4-methyl-benzotriazole (4-TTri) and 5-methyl-benzotriazole (5-TTri) regarding aerobic, denitrifying, sulfate reducing and anaerobic conditions and different nutrients was performed. ASCs, taken from three wastewater treatment plants (WWTP) with different technologies namely a membrane bioreactor (MBR-MH), a one-stage conventional activated sludge plant CAS-E (intermittent nitrification/denitrification) and a two-stage conventional activated sludge plant CAS-M were able to completely eliminate up to 30 mg L^{-1} 5-TTri and BTri under aerobic conditions within 2 - 7 and 21 - 49 days, respectively, but not under denitrifying, sulfate reducing or anaerobic conditions. 4-TTri proved stable at all conditions tested. Significant differences were observed for BTri biodegradation with not-acclimated ASC from MBR-MH with 21 days, CAS-E with 41 days and CAS-M with 49 days. Acclimated ASC removed BTri within 7 days. Furthermore, different carbon and nitrogen concentrations revealed that nitrogen was implicitly required for biodegradation while carbon showed no such effect.

4.1 INTRODUCTION

The xenobiotic compounds benzotriazole (BTri) together with its two derivatives, 5-methyl-benzotriazole (5-TTri) and 4-methyl-benzotriazole (4-TTri), are polar micropollutants commonly used as corrosion inhibitors whenever water comes in contact with metal surfaces. 4- and 5-TTri are commercially available as tolyltriazole mixture (TTri) widely used in metal finishing and as corrosion protection in cooling systems [86, 118]. Among other applications these three benzotriazole compounds, in the following denoted as BTs, are mainly used as anticorrosive in aircraft deicing or breaking fluids [119], in household dishwashing detergents for silver and corrosion protection [63] and as UV filter and/or sun screen agents [102, 120]. With an annual production volume of around 9000 tons in the United States and probably a much higher global production [99, 121], these three benzotriazoles are categorized as 'high production volume chemicals'. Due to their widespread usage, high polarity (logP 1.44 and 1.71 for BTri and TTri, respectively) and thus high water solubility, as well as poor biodegradability, these compounds are found in nearly all aquatic compartments [84, 94, 122, 123]. This includes ground water with up to 1,024 ng L^{-1} [17], river systems with up to 7,997 ng L^{-1} for BTri and 19,396 ng L^{-1} for TTri [53], and the North Sea with up to 40 ng L^{-1} for all benzotriazoles together [109, 110]. Alarmingly high concentrations of up to 128 mg L^{-1} and 198 mg L^{-1} for BTri and TTri, respectively, were recorded for a perched water monitoring well at an international North American airport [111]. At concentrations above 40 mg L^{-1} for BTri and 6 mg L^{-1} for 5-TTri previous work showed significant toxicity in Microtox® tests (*Vibrio fischeri* light emission as toxicity test system, Azur Environmental, CA) [112, 113, 124]. Therefore, BTs are potentially hazardous for the aquatic system [125, 126]. Due to their high efficiency in protecting metal surfaces from corrosion and their relatively low costs, these chemicals are frequently used in industrial and household applications [64, 121]. Unfortunately, even wastewater treatment plants (WWTP) have been shown to be incapable of completely removing these compounds during the treatment process [101, 102]. Therefore, WWTP effluents constitute one major point source for these compounds into the aquatic environment. Several workgroups have reported a variety of removal efficiencies in conventional WWTPs ranging from 13-62% for BTri, 11-74% for 5-TTri, and far less for 4-TTri, which often did not show any removal [34, 49, 85, 127]. Even at defined laboratory conditions, complete biological removal of BTri has until now never been achieved, while 5-TTri was shown to be completely biodegraded under aerobic conditions within 91 days [122]. Long half-lives of BTri and 5-TTri with 114 and 14 days, respectively, make it nearly impossible to achieve such a removal efficiency at true full scale conditions in WWTPs as hydraulic retention times (HRT) are nearly always below 15 h during biological treatment [122]. These poor removal rates in WWTPs explain why these three compounds are ubiquitous in aquatic river systems across Europe that receive treated wastewater [53]. From the 122 tested river systems in the study of Loos et al. (2009), 94% were positive for BTri (mean concentration 493 ng L^{-1}) and 81% for both tolyltriazoles (mean concentration 617 ng L^{-1}).

This study was undertaken to shed more light onto the biodegradation capacity of activated sludge communities (ASC) with respect to three benzotriazoles. Three ASC, taken from three WWTPs with different biological treatment regimes were tested for their ability to biodegrade BTri and TTri under meso- and oligotrophic aerobic as well as anaerobic conditions. Research focused on evaluating the degree of biodegradation at varying concentrations of BTri, 4-TTri, and 5-TTri, given either as sole nutrient source or being supplied together with other nutrients.

4.2 MATERIALS AND METHODS

4.2.1 Chemicals

1-H-benzotriazole (BTri; 99,8 % purity; CAS 95-14-7) and the technical mixture of 4-methyl-benzotriazole (4-TTri; CAS 29878-31-7) and 5-methyl-benzotriazole (5-TTri; CAS 136-85-6), commonly known as tolyltriazole (TTri; 99,5 % purity; CAS 29385-43-1), were provided by Cimachem GmbH (Kirchheimbolanden, Germany). The chemical structures and chromatographic retention times, according to the applied method of the benzotriazoles tested are shown in Figure 4.1. Water solubility of the three benzotriazoles at 20°C is as follows: BTri 19.0 g L^{-1}, 5-TTri 6.0 g L^{-1} and 4-TTri 0.25 g L^{-1}. Water solubility for BTri and 5-TTri was taken from the manufacturer data sheet, 4-TTri experimentally determined by measuring samples with different concentrations. TTri consisted of 40% 4-TTri and 60% 5-TTri. The internal standard 5,6-dimethyl-benzotriazole, sodium carbonate solution, and toluol were purchased from Sigma-Aldrich (Steinheim, Germany), peptone and all other media components from Merck KGaA (Darmstadt, Germany).

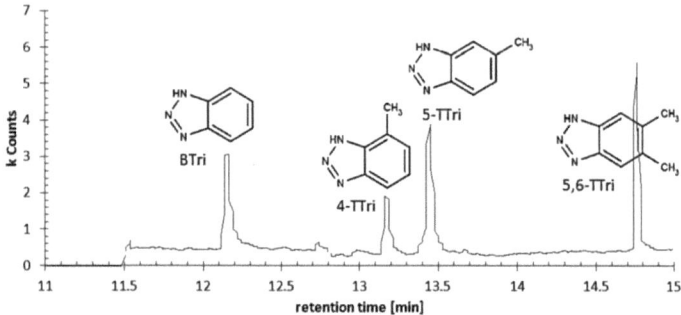

Figure 4.1 – Chemical structures and retention times, according 4.2.4, of the three different benzotriazole compounds and 5,6-methyl-benzotriazole used as internal standard.

4.2.2 Wastewater Treatment Plants' Characteristics and Activated Sludge Sampling

WWTPs were chosen for their differences in population equivalents (PE), treatment processes and type of influent (industrial and/or municipal). Membrane bioreactor MBR-MH receives very little industrial wastewater containing up to 17 µg L^{-1} BTs. The conventional activated sludge plant CAS-E, responsible for treating airport wastewater constituting up to 30% of the total influent volume, receives high concentrations of BTs (up to 91 µg L^{-1}). CAS-M treats wastewater, which receives roughly 50% of its influent volume from industry. It contains high amounts of up to 40 µg L^{-1} BTs due to their use in metal finishing. Main characteristics of the three WWTPs are presented in Table 3.1.

Activated sludge samples (AS) for inoculation, 500 mL each, were collected from the aeration tanks of MBR-MH and CAS-E and from the high load stage aeration tank of CAS-M. All sludge samples were taken as mixed liquor grab samples during aeration to ensure optimal mixing of the sludge community. Fresh digester sludge was only taken from CAS-M.

4.2.3 Experimental Setup

4.2.3.1 Reactor and Control Setups, Nutrient Supply

Biodegradation capabilities of the three ASCs were tested in separate reactor setups. Analyses for BTri, 4-TTri, and 5-TTri were performed individually to investigate different biodegradation patterns.

All experiments lasted for 50 days. Aerobic setups were performed with MSM-PE, MSM-CN, MSM-N, MSM-C and MSM media while denitrifying and sulfate reducing and anaerobic experiments were only conducted with MSM-CN (Table 4.1).

Table 4.1 – Media compositions used for incubation.

media	MSM-PE	MSM-CN	MSM-N	MSM-C	MSM
basic component [g L^{-1}]	adapted from DIN ISO 9888[128]; KH_2PO_4 (0.08), K_2HPO_4 (0.2), Na_2HPO_4 (0.3), $MgSO_4*7\ H_2O$ (0.02), $CaCl_2*2\ H_2O$ (0.04), $FeCl_3*6\ H_2O$ (0.0003); pH 7.4				
additional supplement	peptone (70 mg L^{-1}), DOC: N 2:1	sodium-acetate-C (90 mg L^{-1}), NH_4NO_3-N (9 mg L^{-1})	NH_4NO_3-N (9 mg L^{-1})	sodium-acetate-C (90 mg L^{-1})	-

MSM – mineral salts media

Aerobic MSM-PE and denitrifying and sulfate reducing MSM-CN setups were operated in semi-continuous batch-mode. They were supplied with 70 mg L^{-1} peptone (dissolved organic carbon 28 mg L^{-1}, nitrogen 12 mg L^{-1}) or $NaNO_3$ (0.17 g L^{-1}) or Na_2SO_4 (0.28 g L^{-1}) after each sampling. All other setups were prepared as batch reactors only providing initial nutrients.

Aerobic media were prepared using 500 mL Erlenmeyer flasks with air-permeable aluminum caps filled with 150 mL mineral salt media (MSM) as given in Table 4.1. Aerobic conditions inside the reactors were ensured by using an orbital shaker at 150 rpm (a previous experiment showed this shaking speed to be sufficient to keep the oxygen concentration above 2 mg L^{-1}). Finally, all media solutions were spiked with BTri and TTri at varying concentrations according to Table 4.2.

Table 4.2 – Amounts of BTri, 4-TTri, and 5-TTri used for biodegradation experiments in the different setups. TTri consists of 4- and 5-TTri 40%/60% (w/w), respectively.

origin of inoculum	initial concentration [mg L^{-1}] MSM-PE			initial concentration [mg L^{-1}] MSM-media and denitrifying/ sulfate reducing			initial concentration [mg L^{-1}] anaerobic MSM-CN		
	BTri	4-TTri	5-TTri	BTri	4-TTri	5-TTri	BTri	4-TTri	5-TTri
MBR-MH	21.0	8.8	13.2	0.5	0.2	0.3	10.0	4.0	5.0
CAS-M	34.0	20.8	31.2	0.5	0.2	0.3	10.0	4.0	5.0
CAS-E	22.0	15.2	22.8	0.5	0.2	0.3	10.0	4.0	5.0

Two sets of reactors (see Table 4.4) under denitrifying and sulfate reducing conditions, as they occur in non-aerated AS basins, were created by using 10 L bottles being slowly shaken at 30 rpm.

Biodegradation was tested in MSM-CN media being supplied once a week with $NaNO_3$ and Na_2SO_4.

Anaerobic biodegradation was tested in batch mode, not supplying additional nutrients during the experiment. Media were prepared as described above; bottles sealed with rubber stoppers and aluminum crimps. Setups were performed in MSM-CN media without extra electron acceptors as control or with either $NaNO_3$ (3 mM), Na_2SO_4 (2 mM) or the humic acid model compound anthraquinone-2,6-disulfonate (AQDS, 5 mM) to create similar electron accepting capacities [129] (Table 4.4). Nitrate and sulfate were measured according to DIN 38405 D9-2 with Hach Lange cuvette tests LCK 340 (nitrate) and LCK 353 (Hach Lange GmbH, Duesseldorf, Germany). Cystein-HCl solution (1 mM) was added to all reactors as reducing agent. Anaerobic conditions were maintained by pressurizing the bottles (1.2 bar) with a gas mixture of N_2/CO_2 (80:20, v/v) and controlled by adding resazurin as redox-indicator.

10 mL samples were taken once a week from the experimental reactors after a one-hour sedimentation period to reduce biomass withdrawal. Samples were filled into pre-washed glass bottles and prepared as described in section 2.4.

Duplicate controls were set up in the same manner and consisted of a) abiotic controls without AS and b) sterile controls with AS being autoclaved twice. Controls were operated alike the experimental reactors.

All experiments were carried out in the dark to avoid photolysis. Temperature was kept constant at 20°C (± 2°C) for the duration of the experiments. The pH was controlled in the range of 7 to 8.

4.2.3.2 Inoculation

For inoculation fresh activated sludge was collected from the WWTPs (WWTPs' characteristics in Table 3.1). Sludge was centrifuged (10 min, 4000 g), the supernatants discarded, and the remaining biomass was washed with 1xPBS-Buffer (NaCl (8.0 g L^{-1}), KCl (0.2 g L^{-1}), Na_2HPO_4 (2.7 g L^{-1}), KH_2PO_4 (0.2 g L^{-1})). Repeating the procedure twice ensured to remove wastewater residues and nutrients. Finally, each sludge was adjusted to 5 g L^{-1} mixed liquor suspended solids (MLSS) in MSM-PE reactors and to 3 g L^{-1} MLSS in all other setups.

For anaerobic experiments, fresh digester sludge, collected only from CAS-M, was transferred to the reactors with syringes to maintain anaerobic conditions.

4.2.4 Sample Preparation and Chemical Analyses of BTs

The samples obtained from the reactors were centrifuged (10 min, 8,000 g, 20°C) to remove any cellular debris and biomass, filled into cleaned and sterilized glass bottles and stored in the dark at -20°C before analysis.

Chemical preparation and determination of the BTs' concentration in the liquid samples were performed according to Liu et al. (2011) with the following steps: all samples were derivatized with acetic anhydride, and extracted with toluene. The calibration curve was determined with aqueous solutions containing 0.1, 1.0, 6.3 and 25.0

µg L^{-1} of BTri, 4-TTri, and 5-TTri. These three solutions, high-purity water serving as a blank, and the prepared samples were spiked with an internal standard solution (5,6-dimethyl-benzotriazole in sodium carbonate solution) containing the same concentration as used for the calibration curve. The extracts were analyzed by gas chromatography followed by tandem mass spectrometry (GC/MS-MS) on a Saturn 2200 by Varian (Agilent Technologies Deutschland GmbH, Böblingen, Germany) equipped with an ion trap. Limit of quantification for this setup was 0.01 µg L^{-1}. Compound separation was accomplished on a VF-5ms column from Varian (30 m x 0.25 mm, film thickness 0.25 µm) perfused by helium as the carrier gas at a constant flow rate of 1.5 mL min^{-1}. The temperature profile started at 65°C (held for 4 min), was increased by 12°C min^{-1} to 200 °C, and was finally set to 300°C at a rate of 40°C min^{-1} (held for 6 min). Operation mode of the MS/MS was resonance excitation of the characteristic precursor ions of the analytes and the internal standard. Injection was performed split less ranging from 1.0 to 9.0 µL sample volume (large volume liner, Varian 1079 programmable injector). Compounds were identified by comparing the retention times (within a 2% range of deviation) and product ion spectra with those of the standards (see Figure 4.1).

Blanks were analyzed to check for possible contaminations of the experimental samples and to verify the accuracy of the method itself. Due to the analytical setup (SPE, derivatization) the total error of the method amounts to 15 %.

4.3 RESULTS AND DISCUSSION

4.3.1 Biodegradation of BTri, 4-TTri, and 5-TTri under Aerobic Conditions

4.3.1.1 Benzotriazole (BTri) Biodegradation under Mesotrophic Conditions (MSM-PE)

This experiment was conducted in semi-continuous batch mode by regularly supplying nutrients to emulate pre-treated wastewater nutrient conditions as they appear in aeration basin inflows. Sorption and abiotic processes were negligible as all controls showed no change in the BTs concentrations under the applied conditions proving that observed elimination occurred by biodegradation only. All three ASCs were able to fully eliminate BTri under aerobic conditions (Figure 4.2). Different biodegradation patterns and times, needed for removal, were observed for the three ASCs before acclimation. High BTri concentrations created a strong selective pressure and allowed to link biodegradation not only with ASC origin but also with acclimation to high BTri concentrations (as different initial concentrations were used). A subsequent experiment with acclimated sludges showed a similar biodegradation pattern as all sludges were able to eliminate BTri within 7 days.

Biodegradation occurred fast and was not linear thus biodegradation rates and half-life times could not be calculated as they were significantly changing with the initial concentration. ASC might possess the ability to degrade even higher concentrations up to 30 mg L^{-1} when given enough time for acclimation. Thus, providing biodegradation rates and half-life times for these experiments were not useful for characterization of the ASC removal efficiency.

In semi-continuous batch mode, ASC from MBR-MH totally removed BTri within 21 days followed by CAS-E with 42 days and CAS-M with 49 days. However, BTri elimination was substantially faster than 114 days as stated in a previous publication [122] where no additional nutrients were supplied. Therewith the availability of sufficient nutrients significantly enhanced biodegradation.

The observed BTri biodegradation potential is even more surprising as the fastest biodegrading MBR-MH ASC was confronted with the lowest BTs influent concentrations of the three WWTPs of around 10 µg L^{-1} compared to 18 and 35 µg L^{-1} for CAS-M and CAS-E, respectively (see also Chapter 3). An adaption phase was observed for all three ASC being shortest for MBR-MH with 7 days and longest for CAS-M with almost 45 days. CAS-E and CAS-M are more similar in treatment technique as compared to MBR-MH explaining why CAS-E and CAS-M showed almost the same biodegradation pattern. CAS treatment favors flocculation for faster settlement while MBR treatment allows also planktonic growth of the organisms, as sedimentation is not required. Microorganisms are not pressured towards agglomeration to increase settleability. Therefore nutrient transport via diffusion and uptake is enhanced in MBR treatment while flocks are more limited as nutrient transport is slower due to their larger size in CAS [130, 131].

Another reason for favoring MBR biodegradation performance is sludge age (Table 3.1). Sludge age of MBR-MH was considerably higher than those of CAS-E and CAS–M. Membrane filtration favors slower growing organisms and increases the overall diversity in the system [34, 35, 101]. The possibility that organisms, capable of biodegrading also persistent substances, will be present in the system is considerably higher. That explains why MBR-MH although facing lowest BTs influent concentrations, showed fastest acclimation and biodegradation as MBR-MH ASC are more diverse and already well adapted. CAS-E and CAS-M stage 1 both exhibit lower sludge ages giving the organisms less time to adapt and biodegrade complex substances. As readily degradable nutrient sources are consumed first, time is short for biodegradation of micropollutants such as BTs.

Enhanced transport due to smaller aggregates, i.e. flocks, combined with higher sludge age and therefore better-adapted organisms explain why MBR-MH removes BTri faster compared to CAS. Better adapted organisms and/or enhanced nutrient uptake leads to a removal of BTri within 21 days in MBR-MH. CAS-E and CAS-M have similar treatment technologies and might therefore be supposed to remove BTri in almost the same period. However, CAS-M needed around 8 days more. That extension in time, required for removal, might be due to the higher initial concentration of BTri in CAS-M reactors (Table 4.2) were ASC needed a longer acclimation period. Within seven days all BTri was removed indicating that acclimation to BTri significantly enhanced biodegradation. These findings fit the effect of acclimation already described in literature to enhance biodegradation of e.g. phenol and 4-chlorophenol [132] but have never been shown for BTs before.

Additionally, this observation was supported by an acclimation experiment. Spiking 10 mg L^{-1} BTri into the same reactors again, after complete elimination of the initial BTri concentrations, all three ASC, although taken from three different WWTPs, removed BTri within seven days and no longer showed different biodegradation patterns. Acclimation was found to be one powerful technique to enhance BTri biodegradation efficiency and might help to improve WWTPs removal towards BTri as well as for the other two compounds 5-TTri and 4-TTri. As shown in Chapter 3 BTri is only partially removed during wastewater treatment implicating the need for optimization of the WWTPs removal efficiency.

4.3.1.2 Benzotriazole (BTri) Biodegradation under Oligotrophic Conditions

For testing carbon and nitrogen influences on biodegradation, experiments in MSM-CN, -N, -C and MSM were established (Table 4.1). Low BTri concentrations of 500 µg L^{-1} were used to create environmental relevant conditions as influent concentrations of up to 100 µg L^{-1} were already found [118]. No nutrients were supplied during the experiment to create oligotrophic conditions and a strong pressure towards utilization of BTri as its concentration was too low to act as selective agent compared to 3.1.1 were 10 mg L^{-1} BTri applied a selective pressure. Figure 4.3, compared to Figure 4.2, clearly shows that nutrient supply, nutrient availability and nutrient composition had major effects on biodegradation.

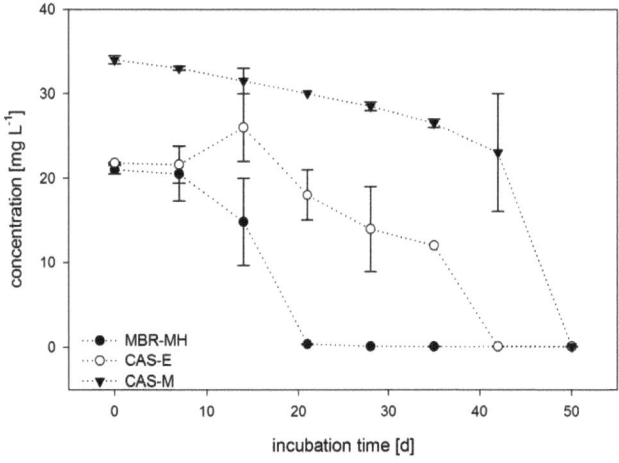

Figure 4.2 – Aerobic BTri biodegradation patterns within the three ASCs. Reactors were operated in semi-continuous batch mode in MSM-PE media. Peptone was added weekly after each sampling. Shown are mean values of BTri concentrations of duplicate experiments (initial concentrations given in Table 4.2) with error bars indicating standard deviations (n=2).

One general observation was that biodegradation under oligotrophic conditions did not differ that much across the three ASC as observed under mesotrophic conditions in semi-continuous batch-mode. MBR-MH showed good biodegradation potential, supporting the above discussed theory of ASC from MBRs being well adapted [101, 127, 133], but in contrast to semi-continuous, mesotrophic setups (Figure 4.2) nutrient composition turned out to be more important than ASC structure. In these batch setups, ASCs activity had more influence on biodegradation, explaining that all three ASC showed similar biodegradation patterns, especially after 50 days of incubation. BTri was removed faster when carbon/nitrogen or nitrogen were provided. MSM-C, i.e. carbon supply only, showed slightly better removal than MSM, i.e. without carbon and nitrogen, but both media approved not useful for effective biodegradation strengthening that nitrogen supply was crucial for effective biodegradation. BTri could obviously not serve as nitrogen source explaining why removal in media containing nitrogen (MSM-N and MSM-CN), occurred at almost the same speed and pattern. Other studies also reported reduced biodegradation efficiency when no additional nutrients, especially nitrogen, were present [122], supporting the above mentioned theory. However, this result is in contrast to other findings where, during biodegradation of pharmaceuticals such as sulfamethoxazole, carbon supply enhanced removal [50, 134]. MSM, containing BTri as sole nutrient source, was not useful for enhancement of biodegradation. BTri alone might not be sufficient to serve as energy source in absence of readily degradable nutrients, especially nitrogen compounds.

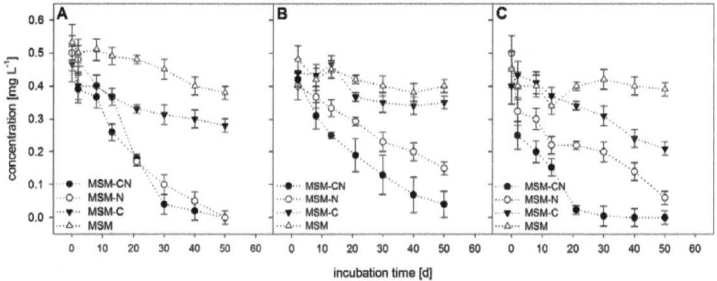

Figure 4.3 – BTri biodegradation with ASC MBR-MH (A), CAS-E (B) and CAS-M (C) in four different media (Table 4.1).
Reactors were operated as batch without any additional nutrient supply. Shown are mean values of BTri concentrations of duplicate experiments (initial concentrations given in Table 4.2) with error bars indicating standard deviations (n=2).

Whether BTri was utilized as carbon source or just transformed into other intermediates is still not known for sure, but nitrogen deficiency significantly hinders biodegradation while carbon deficiency did not seem to have such an effect. Therefore, it was assumed that BTri biodegradation might start by degradation from the benzene ring as the BTri-N could not be used and had to be externally supplied.

4.3.1.3 4-methyl and 5-methyl-benzotriazole (4-TTri, 5-TTri) Biodegradation under Different Nutrient Conditions

4-TTri and 5-TTri showed a completely different biodegradation pattern compared to BTri. Although being closely related structurally, 4-TTri and 5-TTri behaved completely different during biodegradation. After 50 days of incubation the concentration of 4-TTri was still at initial level, regardless which ASC and which media were used. Other studies concerning 4-TTri behavior during wastewater treatment assumed this compound to be relatively stable and only poorly removed [99, 127, 135] but it was shown here for the first time under specific laboratory conditions that 4-TTri was perfectly stable and could not be removed due to biodegradation.

In contrast, complete removal of 5-TTri occurred within maximal 7 days in all tested ASC (shown only for ASC CAS-M) and all media, even in MSM where 5-TTri might serve as sole nutrient source (Table 4.3 and Figure 4.4).

Table 4.3 – Biodegradation of 5-TTri under aerobic conditions with ASC CAS-M. Shown are initial 5-TTri concentration and time needed to remove 99.9% 5-TTri.

used media	initial concentration of 5-TTri [mg L^{-1}]	required time to remove 99.9% 5-TTri [d]
MSM-PE	31.2	≤ 7
MSM-CN	0.3	4-6
MSM-N	0.3	4
MSM-C	0.3	2
MSM	0.3	2

In MSM-PE setups with high 5-TTri concentrations from 13 to 31 mg L^{-1}, within 7 days 5-TTri was removed in all tested ASCs. Thus, ASC tested showed that their biodegradation potential, even at 31 mg L^{-1}, was still not exhausted. All three ASC showed the same pattern and almost same rate showing that organisms are able removing 5-TTri without the need for long acclimation as shown for BTri.

In reactors provided with MSM media and low concentrations of 0.3 mg L^{-1} 5-TTri, this compound was completely eliminated after 2 days for MSM-C and MSM while 4 days were needed for MSM-N and 4-6 days for MSM-CN (Table 4.3 and Figure 4.4). ASC needed slightly more time to remove 5-TTri in MSM-CN as readily degradable nutrients are used up first. Additionally, the fast removal of 5-TTri in MSM could probably mean that 5-TTri can, unlike BTri, serve as sole nutrient source for nitrogen and carbon under the applied experimental conditions. That effect has to be further studied. Slight differences between MSM-N and MSM-C were observed by 5-TTri removal and further experiments are implicitly required to evaluate the impact of nitrogen and/or carbon for the enhancement of 5-TTri biodegradation. 5-TTri elimination in two days takes still too long for an efficient and complete removal during wastewater treatment.

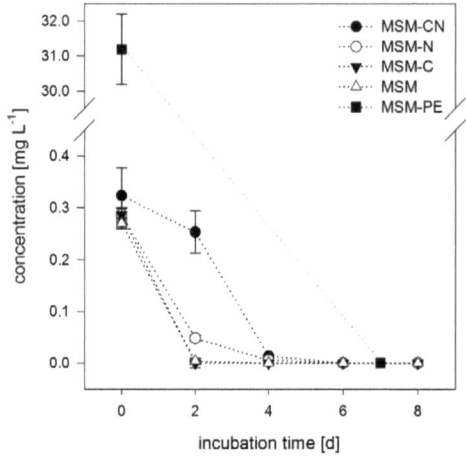

Figure 4.4 – 5-TTri biodegradation with ASC CAS-M and five different media (Table 4.1).
Reactors were operated as semi-continuous batch (MSM-PE) and batch (other media). Shown are mean values of 5-TTri concentrations of duplicate experiments (initial concentrations given in Table 4.2) with error bars indicating standard deviations (n=2).

As the biodegradation behavior of 5-TTri was completely different to that of BTri and 4-TTri, as already observe in chapter 3, it might be likely that the shift of one methyl group to the next adjacent carbon atom makes 5-TTri, in contrast to 4-TTri, a more readily biodegradable substance. Why the position of the methyl group has such a dramatic influence on biodegradation still lacks understanding.

It was also tested if BTri or 4-TTri could be detected during 5-TTri biodegradation as a possible transformation of 5-TTri into either

BTri or 4-TTri as suggested by Liu et al. (2011). However, neither BTri nor 4-TTri was found at any point during 5-TTri biodegradation indicating that during these experiments 5-TTri was not transformed into one of the other two compounds.

4.3.2 Biodegradation Behavior of BTri, 4- and 5-TTri under Low Redox Conditions

Biodegradation was also tested under low redox conditions in presence of nitrate and sulfate as electron acceptors as they also appear in the non-aerated aeration basin and under anaerobic conditions in digesters. AQDS was additionally tested in anaerobic setups as it was shown to foster anaerobic biodegradation for organic compounds in soil [129]. Also for some pharmaceuticals anaerobic biodegradation was shown to be one important way of removal [136].

Table 4.4 – Xenobiotics' biodegradation under low redox conditions. Shown are concentrations given in % from initial value (100%) over incubation time of 30 days for setups with activated sludge and 50 days for digester sludge.

inoculum	media	electron acceptor	redox potential [mV]	remaining conc. [%] after incubation of 30/50 days		
				BTri [%]	5-TTri [%]	4-TTri [%]
activated sludge (CAS-M)	MSM-CN	NaNO$_3$	>-110	96±3	98±1	100±1
		Na$_2$SO$_4$		98±4	107±5	106±5
digester sludge (CAS-M)		-	<-110	99±2	98±2	114±9
		NaNO$_3$		115±10	104±13	103±1
		Na$_2$SO$_4$		96±2	108±8	103±15
		AQDS		104±12	127±11	103±4
abiotic control		-		102±1	104±1	100±3

Thus, two experiments were conducted with either activated sludge not being aerated and with digester sludge in sealed flasks under nitrogen atmosphere. As MSM-CN media was found to show best biodegradation in aerobic batch experiments, these setups were conducted solely with MSM-CN. By monitoring nitrate and sulfate concentrations, the sludge biomass was found to be active as either denitrification or sulfate reduction occurred. As given in Table 4.4, neither of the BTs was biodegraded under these conditions within the experimental period. Compared to Liu et al. (2011), where slight anaerobic biodegradation occurred within 50 days, the initial concentration of BTri, 4- and 5-TTri remained constant after 50 days of incubation under the tested conditions. That indicates aerobic biodegradation to be the major mechanism to biologically removing these BTs from the aquatic system. This is corroborated by the findings of Liu et al. (2011).

4.4 CONCLUSIONS

Activated sludge communities, collected from three wastewater treatment plants with different treatment regimes, were capable of eliminating BTri and 5-TTri, but not 4-TTri, under aerobic, mesophilic conditions in the dark. 4-TTri was stable towards biodegradation. Experiments with different nutrients showed that nitrogen availability is more important than carbon supply for BTri biodegradation while 5-TTri was removed regardless the nutrients. Fastest biodegradation occurred for 5-TTri followed by BTri. BTri showed sludge specific biodegradation patterns but, after sludges acclimation, was removed with the same pattern.

Under denitrifying and sulfate reducing as well as under anaerobic conditions in presence of different electron acceptors, none of the three compounds showed a biological removal. Thus, presumably aerobic biodegradation is the major removal mechanism in aquatic systems.

Chapter 5

Enhancement of 5-TTri Biodegradation with Activated Sludge Communities by Means of Acclimation and Nutrient Supply

The corrosion inhibitor 5-tolyltriazole (5-TTri) is only partly eliminated during wastewater treatment, can have a detrimental impact on aquatic systems and thus implies an acute need to reduce its concentration in wastewater treatment plants (WWTP) effluents. The focus of this study is to enhance the biodegradation of 5-TTri through acclimation as well as nutrient supply. Activated sludge communities (ASC), capable of biodegrading 5-TTri, were used to inoculate different generations resulting in nine ASC generations. Generation two showed a lag phase of five days, after which rapid biodegradation occurred. Generations four to nine were immediately able to utilize 5-TTri after inoculation, with rates ranging from 3.3 to 5.2 mg L^{-1} d^{-1}. Additionally, the supernatant from centrifuged AS was used to simulate wastewater nutrient conditions. Sludge supernatant (SS) enhanced biodegradation and showed that 5-TTri removal might be dependent on nitrogen supply and presumably begins by benzene ring cleavage. Subsequent experiments, performed with three nitrogen species also proved sufficient in enhancing biodegradation. All nitrogen species showed similar biodegradation enhancement and enabled the ASC to utilize 5-TTri without the need for acclimation.

5.1 INTRODUCTION

The xenobiotic compound 5-methyl-benzotriazole (5-TTri) is a commonly used corrosion inhibitor for protection of metal surfaces. 5-TTri is, in combination with 4-TTri, commercially available as tolyltriazole (TTri) and widely used in metal finishing and in cooling systems [86, 118]. 5-TTri is also used as anticorrosive in aircraft deicing or breaking fluids [119], in household dishwashing detergents for silver and corrosion protection [63, 83] and as UV filter and/or sun screen agents [65, 90]. Due to its high production volume, widespread usage, high polarity (logD 1.71), good water solubility and limited biodegradability, 5-TTri is almost omnipresent in aquatic compartments [89, 94, 109, 137]. In river systems tolyltriazole was detected up to 134 µg L^{-1} for TTri [53] but no concentrations for 5-TTri itself were available.

5-TTri showed toxic effects above concentrations of 6.0 mg L^{-1} as reported in a previous study with Microtox® tests (*Vibrio fischeri* light emission as toxicity test system) [113, 138]. Chronic effects of 5-TTri occurred already at concentrations of 0.40 mg L^{-1} and showed adverse effects in the aquatic organism *Daphnia galeata* [112]. Therefore, 5-TTri is generally regarded as potentially hazardous for aquatic systems [114] implying the need to reduce influent concentrations and optimize the removal efficiency of this compound. Especially wastewater treatment plants (WWTP) were, besides small diffuse entry paths from road runoffs, found to be the major point source for 5-TTri into the aquatic environment [63, 82, 85] as 5-TTri was often incompletely removed during treatment [26, 101, 102]. In literature various laboratory biodegradation experiments examining the biodegradation behavior of 5-TTri are described. Such experiments showed that 5-TTri can be eliminated e.g. by microbial communities from activated sludge and aquifer material [89, 90]. However, these studies were performed without microbial activated sludge community (ASC) acclimation or optimal nutrient availability. Therefore, information on 5-TTri removal enhancement with respect to biomass acclimation and nutrient composition is implicitly required.

For that purpose, the main objectives of this study were to evaluate efficient conditions to optimize aerobic 5-TTri biodegradation. This was accomplished by performing several acclimation steps to enhance ASCs' 5-TTri biodegradation. Additionally, the effect of nutrient availability was evaluated by applying a sludge supernatant derived from activated sludge to simulate wastewater nutrient conditions as well as by additional supply of nitrogen containing compounds. Therefore, biodegradation setups, continuously sampled, are used for in depth-studies on 5-TTri biodegradation.

5.2 MATERIALS AND METHODS

5.2.1 Chemicals

5-methyl-benzotriazole (5-TTri; CAS 136-85-6) was purchased from Sigma-Aldrich (Steinheim, Germany), all other media components were obtained from Merck KGaA (Darmstadt, Germany). Milli-Q water was prepared with a Milli-Q system (Millipore, Billericia, MA, USA).

5.2.2 Experimental Setup for Biodegradation

Biodegradation was tested in different setups carried out in three media R2A-UV, MSM-SS and MSM used for different experiments (Table 5.1). All setups were supplied with 20 mg L^{-1} 5-TTri and were prepared in 100 mL glass bottles filled with 20 mL of media and covered with air-permeable aluminum caps. Aerobic conditions inside the reactors were ensured by shaking at 150 rpm. No additional nutrients were supplied during the experiment, i.e. the experiments were operated in batch mode.

Table 5.1 – Media compositions, carbon-nitrogen ratio, nutrient applications and experimental setups.

media	components [g L^{-1}]	DOC:N ratio (DOC: N [mg L^{-1}])	application	used for experiment
R2A-UV (R2A media for UV-AM)	casein peptone (1.0), glucose (0.5), potassium phosphate (0.3), soluble starch (0.3)	7:1 (880: 120)	optimal growth conditions, non-selective	pre-evaluation of biodegradation potential
MSM (mineral salt media)	as MSM-CN, without sodium acetate and NH_4NO_3	-	selection of bacteria, utilization of 5-TTri as sole C and N source	acclimation and specific nitrogen supply
MSM-SS	MSM media supplied with 10% sludge supernatant	2.7:1 (162: 62)	simulation of wastewater conditions	supply of sludge supernatant

- Hoagland trace elements (0.1 mL L^{-1}) were added to all media
- pH was adjusted to 7.4 in all media

All experiments were run for a maximum of 10 days and operated in the dark to avoid photolysis. Temperature was kept constant at

20°C (± 2°C) for the duration of the experiments. The pH was controlled to be in the range of 7 to 8. Sterile setups (media and twice autoclaved AS) and abiotic controls (media without biomass) were treated in the same manner as biodegradation experiments.

5.2.3 Activated Sludge Inoculum

Original activated sludge (AS) was taken from stage 1 of a 2-stage municipal conventional activated sludge plant (CAS-M) treating 1 million population equivalents. Stage 1 is the high load stage with a food to microorganism ratio of 0.64 kg BOD_5 kg^{-1} $MLSS^{-1}$. The influent consists of municipal and industrial wastewater (1:1, v/v). 500 mL AS was collected in pre-cleaned 1L glass bottles stored at 4°C and was used within 24 h for reactor inoculation.

Biodegradation potential was pre-evaluated in 150 mL R2A-UV media spiked with 20 mg L^{-1} 5-TTri to apply a high selective pressure. The setups were inoculated to a total biomass concentration of 3 g L^{-1} mixed liquor suspended solids. After biodegradation occurred, the experiment was stopped and 1.0 mL of mixed biomass was used for inoculation of the subsequent acclimation and nitrogen addition experiments. For sludge supernatant experiments inoculation see 2.4.

5.2.4 Acclimation Procedure

Acclimation to 5-TTri was achieved in MSM over several generations from one to nine in the following manner: First generation reactor was inoculated with 1.0 mL pre-evaluated, acclimated AS and was kept running until biodegradation occurred (after 15 days). The experiment was stopped and 1.0 mL of the reactor suspended biomass was used for inoculation of a subsequent experiment termed generation two. After observing biodegradation the generation two experiment was stopped and the third generation reactor was prepared in the same manner as the second-generation one. These setups were repeated until generation nine, obtaining a highly selected biomass for 5-TTri biodegradation.

Biomass from all generations was harvested by centrifugation (20 min, 10,000 g, 4°C), the pellet resuspended in PBS and stored at 4°C. 1.0 mL of this biomass was used to inoculate the sludge supernatant experiments.

5.2.5 Specific Nutrient Supply

A) Sludge Supernatant Experiments

A supplement derived from AS was tested and prepared by autoclaving (20 min, 121°C) AS twice followed by centrifugation (10 min, 10,000 g) and filtration of the supernatant at 0.45 μm to remove all solid particles. The obtained sludge supernatant (SS) was added at a total amount of 10% (v/v) to MSM media (Table 5.1) producing MSM-SS. These setups were inoculated with 1.0 mL reactor suspended solids obtained from generations 2 to 6 from the acclimation experiment.

B) Nitrogen Addition Experiments

Setups were prepared by adding the three different nitrogen species NH_4NO_3, $NaNO_3$ and NH_4Cl at different nitrogen concentrations to MSM media. Total N-concentrations of 25, 50,

100, 250 and 500 mg L^{-1} were used and supplied together with 20 mg L^{-1} 5-TTri.

5.2.6 Detection of Biodegradation

200 μL supernatant, taken after 30 min sedimentation from the setups, were used for UV-absorbance measurements (UV-AM) as described in Chapter 2 with the following changes applied: Calibration was accomplished with 5, 10, 15 and 20 mg L^{-1} 5-TTri in high-purity water and the used media to evaluate measurement reliability and background absorbance. Measurements were accomplished in 96 well UV-star plates from Greiner Bio-One (Greiner Bio-One GmbH, Frickenhausen, Germany) in 200 μL volumes and analyzed with an automated plate reader (EnSpire® Multimode Plate Reader, Perkin Elmer, Rodgau, Germany). For each measurement a 5-TTri blank (media with SMX but without organisms) was measured to detect changes over time as well as a blank (media without 5-TTri) to detect background absorbance. 5-TTri removal was calculated by using the corresponding absorbance values at a wavelength of 262 nm. Decreasing values indicated biodegradation. Removal is given in C/C_0 by dividing the initial absorbance value (C_0) by the values measured after a specific incubation time (C).

5.3 RESULTS AND DISCUSSION

5.3.1 Activated Sludge Acclimation at Low-Nutrient Conditions

Sterile setups approved sorption onto biomass to be negligible and abiotic reactor setups showed 5-TTri to be stable as photodegradation was excluded by performing all experiments in the dark.

The pre-evaluation experiment performed in R2A-UV media inoculated with original AS obtained right from the WWTP showed that the ASC was capable to biodegrade 20 mg L^{-1} 5-TTri. In order to cultivate as many organisms as possible the complex nutrient composition of R2A was used to foster growth of aquatic organisms [139]. Usage of 20 mg L^{-1} 5-TTri was assumed to apply already a strong selective pressure towards organisms able to utilize 5-TTri. As biodegradation occurred, 1.0 mL of reactor suspended biomass was used to inoculate the following acclimation experiment.

All following acclimation setups were performed in MSM media to create oligotrophic conditions being assumed to force the ASC towards faster 5-TTri consumption. First generation ASC setup was discarded as it contained sludge residues that interfered with UV-AM. Therefore, a comparable and sensitive measurement with respect to the subsequent generations was not possible as the sludge created a strong fluctuation and background absorbance that could not be controlled. However, it was possible to detect the general biodegradation trend in the first generation setup revealing that this generation behaved similar to the ASC used for the pre-evaluation test. Comparable and reliable measurements were possible from generation two onwards. This is the reason why the acclimation experiment was started from that generation (Figure 5.1). The 2nd generation ASC was found efficient to degrade 5-TTri but required around 5 days, where no biodegradation occurred, to acclimate to the low nutrient conditions in MSM and the utilization of 5-TTri as sole nutrient source (Figure 5.1 and Table 5.2). This lag phase might be required because the metabolic processes needed for 5-TTri biodegradation had to be established in the cells, a scenario being also observed by other studies investigating biodegradation of phenolic compounds [140, 141]. However, once the cells' metabolism was able to utilize 5-TTri, biodegradation was found to be very fast with a removal rate of 4 mg L^{-1} d^{-1}.

Table 5.2 – Biodegradation rates taken as the slope from the linear fitted lines of the ASC derived from the acclimation and nutrient experiments.

ASC generation	experimental biodegradation rate [mg L^{-1} d^{-1}]		
	acclimation (Figure 5.1)	supply of sludge supernatant (Table 5.1)	
		addition of SS (MSM-SS)	control group (MSM)
2nd	4.0	3.2	< 1.0
3rd	5.0	4.4	< 1.0
4th	3.3	4.3	2.0
5th	4.4	5.0	2.2
6th	4.7	4.2	2.2
8th	5.1	-	-
9th	5.2	-	-

The 3rd generation ASC originated out of the second, showed a similar behavior but with a reduced lag phase of four days and an increased biodegradation rate of around 5 mg L^{-1} d^{-1}. As a lag phase was still required for the ASC to adapt to the oligotrophic conditions, it was clear that the organisms were still not fully acclimated to solely using 5-TTri as nutrient.

The 4th generation ASC was characterized by a significantly improved 5-TTri biodegradation capacity. A lag phase was not observed any more indicating that the cells' metabolism was completely adapted to utilizing 5-TTri as nutrient source but the lower biodegradation rate of 3.3 mg L^{-1} d^{-1} compared to the 3rd generation ASC indicated that there is still optimization required regarding the cells' biodegradation potential. Although the total time needed to achieve an almost complete biodegradation of 20 mg L^{-1} 5-TTri was with just around six days significantly less compared to the former generations, subsequent experiments were performed to achieve the ASC full biodegradation potential.

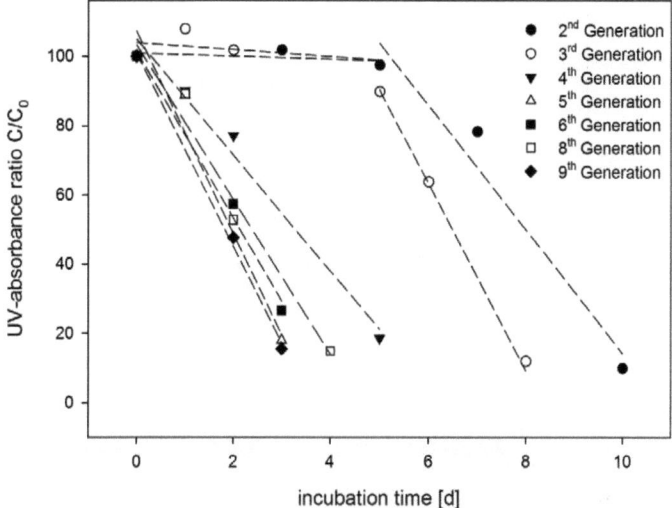

Figure 5.1 – 5-TTri removal monitored by UV-absorbance measurements. Acclimation was achieved in MSM over nine generations. The biomass of the precursor generation was used as inoculum for the next generation. Dashed lines represent linear removals calculated by a linear function (R^2 values were in the range from 0.71 to 0.98).

All following ASC generations did not show any lag phase and were able to utilize 5-TTri immediately after inoculation without the need for adaption. The 5th generation ASC again showed a better removal rate compared to the previous generation with a biodegradation rate of around 4.4 mg L^{-1} d^{-1} and a total removal time ranging from 4 to 5 days. From the 5th generation onwards no significant further improvements regarding biodegradation rate could be achieved. The following generations showed a similar removal rate ranging from 4.7 to 5.2 mg L^{-1} d^{-1} with its maximum in ASC generations eight and nine, indicating no further improvements due to adaption. These biodegradation rates were significantly higher than the one of 0.05 mg L^1 d^{-1} reported in a similar study with an initial concentration of 1.0 mg L^{-1} [90]. The higher initial concentration used here might explain the higher biodegradation rates as the selective pressure towards utilization of 5-TTri was far higher.

Generally, ASC generations two to nine achieved a significant improvement regarding their lag phase, where no biodegradation occurred, whereas the removal rate did not change to the same extent as it was found to be in a narrow range from 3.3 to 5.2 mg L^{-1} d^{-1}. Other acclimation studies focusing on biodegradation of phenolic and other xenobiotic compounds by ASC showed a similar biodegradation strategy with respect to acclimation of the inoculum and biodegradation rate [132, 142, 143]. The present study showed that 5-TTri was easily removed by the ASC when time for acclimation was given. The organisms already present in activated sludge are able to utilize 5-TTri as sole nutrient source by switching on the required metabolic processes. Nevertheless, the conditions applied during that experiment might not be stringent enough to specifically cultivating a microbial community consisting solely of organisms able to utilize 5-TTri. However, the acclimation conditions were adequate to enhance 5-TTri biodegrading ASC metabolism as indicated by the completely omitted lag phase with no biodegradation from the third to the fourth ASC generation. Additionally, this showed the ASC' ability to completely adapting to 5-TTri utilization as sole nutrient within a short time. Further research has to show how long the immediate 5-TTri utilization ability, i.e. no lag phase required, will remain within the ASC if no 5-TTri is present and if low 5-TTri concentrations, as they are found in WWTPs, will be enough to acquire this competence.

5.3.2 5-TTri Biodegradation Enhancement by Nutrient Supply

To screen for further optimization possibilities derived from the utilization of specific nutrients, the following experiments were performed to evaluate the ASC behavior in presence of different substrates.

5.3.2.1 Effect of Sludge Supernatant on 5-TTri Biodegradation

Sludge supernatant (SS) was tested as a specific nutrient to simulate wastewater nutrient concentration and composition. Two separate experiments consisting of five different setups each were initially inoculated with 1.0 mL reactor suspended biomass of biodegrading ASC from generations 2 to 6 obtained from the previous experiment (Figure 5.2) resulting in 10 setups termed again generation 2 to 6. One experiment was performed with MSM media (control setups) while the second was additionally supplied with SS (MSM-SS setups).

Chapter 5 – 5-Tolyltriazole Acclimation

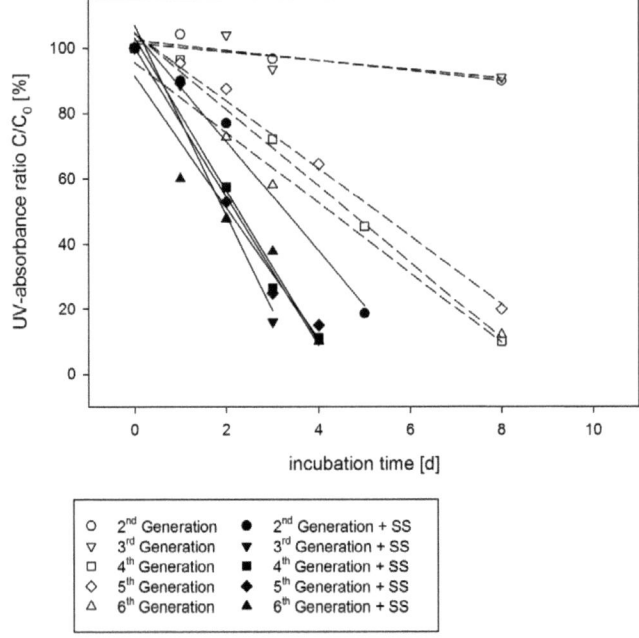

Figure 5.2 – 5-TTri removal in sludge supernatant supplied setups monitored by UV-absorbance measurements.
Sludge supernatant supplied (MSM-SS, solid lines, black symbols) and non-supplied (MSM, dashed lines, white symbols) setups with dashed lines representing linear removal calculated by a linear function (r^2 values were in the range from 0.95 to 0.97). Inoculation was accomplished with 1.0 mL reactor suspended biomass from ASC generations two to six obtained from the previous acclimation experiment (Figure 5.1 and subchapter 5.3.1).

The control setups generations two and three without SS showed a similar behavior as the acclimation experiment regarding lag phase (Figure 5.2, white symbols). The first two ASC generations showed almost no removal in eight days while the remaining generations showed 5-TTri degradation with reduced rates around 2.2 mg L^{-1} d^{-1} and no lag phase as they immediately started biodegradation after inoculation (Table 5.2). The control experiment additionally showed that acclimation towards 5-TTri can be lost when 5-TTri is not continuously present in the system. Biodegradation rates were lower and lag phases for ASC generation two and three longer as compared to the previous acclimation experiments that might be attributed to the biomass storage conditions. Biomass used for inoculation of this experiment was derived from the previous one. To be able to inoculate all reactors at the same time, biomass was stored at 4°C for four weeks. Therefore, inoculation was performed with not actively biodegrading biomass at the start of this experiment and the far lower biodegradation rates demonstrated that storage conditions significantly influenced 5-TTri biodegradation efficiency.

In contrast to these results, the five reactor setups supplied with SS showed a completely different behavior (Figure 5.2, black symbols). All five ASC generations showed similarly high biodegradation potential not requiring any acclimation and achieved degradation rates ranging from 3.2 to 5.0 mg L^{-1} d^{-1}. Even ASC generations two and three immediately started biodegradation after inoculation and did not require any lag phase for acclimation as was observed for the control group. The addition of SS enabled the ASC to readily biodegrade 5-TTri without the need for acclimation after a four weeks storage at 4°C. High biodegradation rates for all five ASC generations indicated that SS directly pushed cellular metabolic processes to the limit as already the 2nd generation showed a three times higher biodegradation rate than the corresponding generation from the control setup.

Chemical analysis of SS revealed a carbon concentration of 162 mg L^{-1} and a nitrogen content of 62 mg L^{-1} i.e. a C:N ratio of 2.7:1. As just 10% (v/v) SS were added to the setups a very low carbon and nitrogen concentration was supplied which was improbable to enhance 5-TTri biodegradation. Seemingly, SS composition rather than concentration affected biodegradation. Sludge supernatant specific compounds, e.g. metabolic precursor substrates, amino acids or other small molecules, might be present in the SS in slight traces only but were required by organisms to directly foster metabolic processes or increase the cells' metabolic activity towards utilization of 5-TTri. These "unknown" substances, contained in media derived from nature, were also found to be important for growth of marine bacteria [144]. Simulating natural wastewater conditions was therefore used as a successful alternative approach to foster growth of "uncultivable cultures" as was also described by other studies focusing on the enhancement and isolation of unknown species from various environments [145-147].

It seemed likely that the carbon-nitrogen ratio might play a major role for optimization of 5-TTri biodegradation as was already found for obtaining optimal growth conditions in a microbial fuel cell [148]. Sludge supernatant supply was found to be an effective way to strongly improve 5-TTri biodegradation probably due to its "natural" composition.

5.3.2.2 Effect of Nitrogen on 5-TTri Biodegradation

As it was found to be likely that nitrogen supply plays a central role for 5-TTri biodegradation, experiments with three different nitrogen species in varying concentrations were conducted. NH_4NO_3, NH_4Cl and $NaNO_3$ were added to MSM media, containing no additional carbon source, in the manner that nitrogen-N concentrations of 0 (= control setup), 25, 50, 100, 250 and 500 mg L^{-1} were achieved. Each reactor was inoculated with reactor suspended biomass obtained straight from the pre-evaluation biodegradation experiment that already biodegraded 5-TTri. For this experiment, unlike the acclimation experiment (see 5.3.1), pre-evaluation sludge was used for inoculation as only screening for the general biodegradation trend was the aim of this experiment and a high sensitivity was not required. In addition, the effect of nitrogen on not yet fully acclimated ASCs was observed. Therefore, inoculation was performed with sludge that was not specifically adapted to 5-TTri biodegradation over several generations.

Figure 5.3 A shows the ASC biodegradation behavior in presence of increasing $NaNO_3$ concentrations. Best biodegradation compared to the control setup was observed in presence of 50 and 100 mg L^{-1} additional nitrogen. The other three concentrations also provoked a very fast removal that was not as strong as observed with 50 and 100 mg L^{-1} nitrogen-N. The addition of NH_4Cl also showed a significant biodegradation improvement being strongest at 25 mg L^{-1} additional nitrogen supplement in comparison to the control setup (Figure 5.3 B). The four higher concentrations of ammonium chloride slightly reduced biodegradation efficiency compared to the 25 mg L^{-1} setup.

The same general effect was observed for NH_4NO_3 (Figure 5.3 C). Biodegradation was significantly enhanced at all concentrations as compared to the control setup. Best biodegradation was found for concentrations of 25, 50 and 250 mg L^{-1} while 100 and 500 mg L^{-1} showed a slightly reduced biodegradation capacity.

In general, all nitrogen-supplied setups eliminated 5-TTri considerably faster than those without N-supply and were even faster as compared to ASC generation 2 from the acclimation experiment (Figure 5.1). Neither nitrogen species composition nor concentration was found to specifically affect biodegradation potential as all reactor setups were able to remove 5-TTri significantly better than the control without additional nitrogen-N.

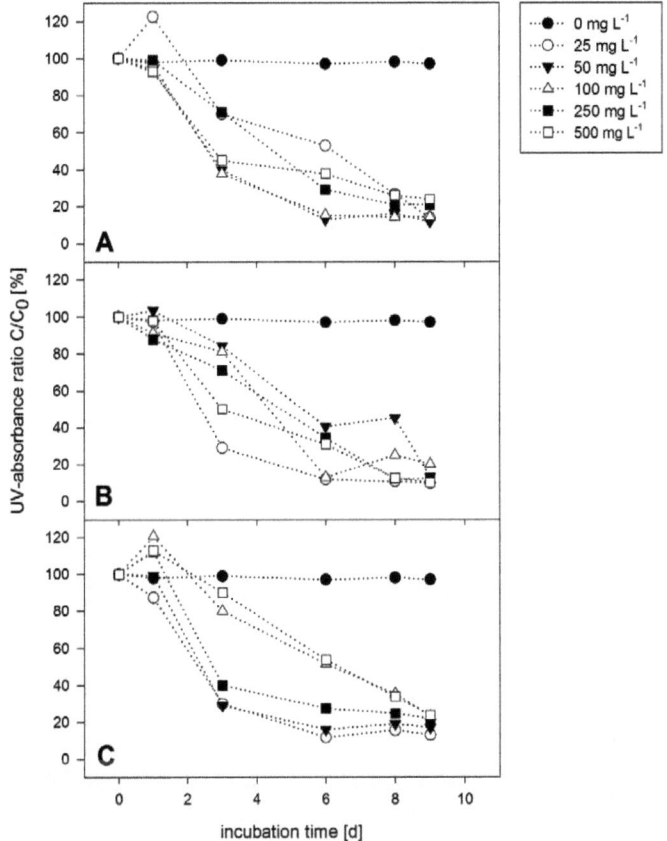

Figure 5.3 A, B, C – 5-TTri removal monitored by UV-absorbance measurements. MSM control setup without nitrogen supply and setups spiked with A) NaNO$_3$, B) NH$_4$Cl and C) NH$_4$NO$_3$ in varying concentrations ranging from 25 to 500 mg L^{-1}. Inoculation was accomplished with 1.0 mL suspended biomass obtained from the pre-evaluation experiment.

Therefore, it seems likely that nitrogen presence is required to foster efficient biodegradation. Moreover, low concentrations of rather unspecific nitrogen compounds were already sufficient, probably due to carbon limitation in this experiment. In wastewater, carbon and nitrogen are present simultaneously enabling the ASC to use these nutrients to produce biomass. Without carbon, nitrogen is mainly used for cell metabolic processes and therefore already low nitrogen concentrations were sufficient to enhance 5-TTri biodegradation. These results suggest that the activity of ASC towards utilization of 5-TTri is strongly related to nitrogen availability rather than carbon supply. This is obvious when comparing the biodegradation with SS supply (Figure 5.2) with the results observed during nitrogen supply. Nitrogen-supplied ASC showed a similarly high 5-TTri removal as the SS-supplied ASC that also contain carbon sources. Therefore ASCs might be able to obtain carbon directly from 5-TTri without the need for an additional carbon supply. Biodegradation of 5-TTri probably start right from the attack of the benzene ring rather than the heterocyclic nitrogen-containing ring as suggested by a previous study [90]. The benzene ring could serve as carbon source as it is known that such ring structures, e.g. phenol, are readily biodegraded [149]. Heterocyclic nitrogen containing compounds are often more persistent to biodegradation, especially the 1,2,3-triazole structure contained in 5-TTri that is often found in fungicides [150, 151]. Therefore nitrogen supply significantly enhanced biodegradation as nitrogen could not be regained from 5-TTri.

5.4 CONCLUSION

Stepwise acclimating an ASC significantly enhanced 5-TTri biodegradation. Subsequent experiments, using the actively biodegrading biomass from the precursor experiment as inoculum, resulted in a multi-generation setup. 5-TTri biodegradation was enhanced three times with a biodegradation rate of around 4 mg L^{-1} d^{-1}. This multi-generation acclimation method was sufficient in reducing acclimation time and simultaneously increased 5-TTri consumption rate.

In addition, sludge supernatant (SS), derived from activated sludge, was found to significantly enhance 5-TTri biodegradation. Setups supplied with SS did not require any time for acclimation and showed high and consistent removal rates around 5 mg L^{-1} d^{-1}. Seemingly, SS contained specific substances that fostered biodegradation by increasing ASC activity and enhanced the cells' metabolic processes rather than enabling biomass growth.

Subsequent experiments with three nitrogen-containing compounds demonstrated nitrogen to be a crucial factor for biodegradation. In presence of nitrogen, independent of the nitrogen species and applied concentrations, biodegradation was almost as fast as in setups with SS indicating that 5-TTri might serve as a carbon source rather than a nitrogen source. Therefore, 5-TTri biodegradation might start from degradation of the benzene ring. Generally, acclimation of ASC and sufficient nitrogen supply effectively enhanced 5-TTri biodegradation.

Chapter 6

Comparative Analysis of Metagenomic Data and DGGE Results Regarding 5-Tolyltriazole Biodegrading Activated Sludge Communities

5-tolyltriazole (5-TTri), only partly eliminated during wastewater treatment, can have a detrimental impact on aquatic systems. Reducing the concentration of 5-TTri in the effluents of wastewater treatment plants (WWTP) is vitally required. Therefore, this study focused on the characterization of 5-TTri biodegrading activated sludge communities (ASC) by means of denaturing gradient gel electrophoresis (DGGE) compared with metagenomic analysis of the whole ASC. DGGE analysis revealed four dominant species Aminobacter spp., Flavobacterium spp., Hydrogenophaga spp. and Pseudomonas spp.. Metagenomic analysis by mapping de novo assembled sequences against the NCBI database showed different prominent species according the media: in MSM-CN (acetate/ammonium nitrate-supplemented) medium Mesorhizobium spp. and Hydrogenophaga spp., in MSM-SS (i.e. sludge supernatant-supplemented medium) Acidovorax spp., Hydrogenophaga spp. and Pseudomonas spp.. In general, eighth generation setups were better adapted to 5-TTri biodegradation than first generation setups.

6.1 INTRODUCTION

The xenobiotic compound 5-methyl-benzotriazole (5-TTri), commercially available as mixture tolyltriazole, is a commonly used corrosion inhibitor for protection of various surfaces. Tolyltriazole, consisting of 5-TTri and 4-TTri in varying proportions, is widely used in metal finishing and cooling systems [86, 118], as anticorrosive additive in aircraft deicing or breaking fluids [119], in household dishwashing detergents for silver and corrosion protection [63, 83] and as UV filter [102, 120]. 5-TTri, polar and good water-soluble, is widely used what results in an almost omnipresent occurrence in aquatic compartments [84, 94, 122, 123]. Wastewater treatment plants (WWTPs) are incapable of completely removing 5-TTri during treatment [101, 102]. Besides diffuse entry paths from road runoffs [17, 82, 85], WWTPs are the major influent pathways. Thus, optimization is implicitly required to improve the treatment process with respect to enhanced 5-TTri removal. Due to its discharge from WWTPs the mixture tolyltriazole (TTri) was detected with concentrations up to 134 µg L^{-1} in river systems [53] and up to 0.51 µg L^{-1} in groundwater that is used for drinking water generation [17].

These results are all the more problematic as 5-TTri already has shown acute toxicity in concentrations above 6.0 mg L^{-1} as reported by a previous study with Microtox® tests (*Vibrio fischeri* light emission as toxicity test system) [113, 124]. Chronic effects were already observed at even lower concentrations of 0.40 mg L^{-1} where 5-TTri showed adverse effects in the aquatic organism *Daphnia galeata* [112]. 5-TTri is generally regarded as potentially hazardous for aquatic systems [114] which implies the need to enhance the removal efficiency of this compound in WWTP systems. Laboratory biodegradation experiments, being conducted to examine the behavior of 5-TTri, showed that 5-TTri can be eliminated biologically in activated sludge systems [90] and also within aquifer material [89]. However, up to now no studies are available describing the composition of the activated sludge community (ASC) that is capable of 5-TTri biodegradation. For a successful optimization of the removal efficiency of activated sludge systems, information on community structure and participating organisms is implicitly required. This knowledge might be also useful to understand why 5-TTri, although chemically closely related to 4-TTri, is far better biodegraded and might ultimately be used to also improve biological 4-TTri removal.

To address that lack of knowledge, this study evaluates the community structure of two acclimated 5-TTri biodegrading activated sludge-derived communities by denaturing gradient gel electrophoresis (DGGE) and compares the results with a metagenomic and metatranscriptomic analysis approach. By doing so information about the ASC composition (who is in there) and also about the bacteria-specific activity was retrieved allowing a detailed view on the abundance of ASC organisms that might play a crucial role in 5-TTri biodegradation.

6.2 MATERIALS AND METHODS

6.2.1 Chemicals and Reagents

5-methyl-benzotriazole (5-TTri; CAS 136-85-6) was purchased from Sigma-Aldrich (Steinheim, Germany), all other media components were from Merck KGaA (Darmstadt, Germany). DNA and RNA cleanup kits were from Qiagen GmbH (Hilden, Germany).

6.2.2 Experimental Setup for Biodegradation

Biodegradation was tested in separate setups with the two media MSM-CN, containing only specific and readily degradable carbon and nitrogen sources, and MSM-SS, containing unspecific, complex nutrients (Table 6.1). All setups were supplied with 20 mg L^{-1} 5-TTri. Aerobic media were prepared using 1000 mL Erlenmeyer flasks with air-permeable aluminum caps filled with 400 mL media. Aerobic conditions inside the reactors were ensured by shaking at 150 rpm. All experiments were performed in batch mode, i.e. no additional nutrients were supplied during the experiment.

Table 6.1 – Compositions, DOC to N ratios and applications of the used media for biodegradation experiments.

media	components [g L^{-1}]	DOC:N ratio (DOC:N [mg L^{-1}])	application
MSM-CN (media with C and N)	KH_2PO_4 (0.08), K_2HPO_4 (0.2), Na_2HPO_4 (0.3), $MgSO_4 \cdot 7 H_2O$ (0.02), $CaCl_2 \cdot 2 H_2O$ (0.04), $FeCl_3 \cdot 6 H_2O$ (0.0003), sodium acetate (0.3) and NH_4NO_3 (0.0075)	33:1 (90:2.7)	co-metabolism
MSM (mineral salt media)	as MSM-CN, without sodium acetate and NH_4NO_3	-	selection of bacteria, growth with SMX as sole C and N source
MSM-SS	MSM media supplied with 10% sludge supernatant	2.7:1 (162:62)	mimic natural wastewater conditions, co-metabolism

- Hoagland trace elements (0.1 mL L^{-1}) were added to all media
- pH was adjusted to 7.4 in all media

Experiments for metagenome, metatranscriptome and DGGE analysis were run for 22 days. Reactors were inoculated with 1.0 mL of generation one and eight of acclimated biodegrading reactors. 5-TTri removal was monitored by UV-absorbance measurements (UV-AM) at 262 nm and was calculated by using the corresponding absorbance values (see Chapter 2.2.4.1 for a detailed description). Decreasing values indicated biodegradation. Removal is given in C/C_0 by dividing the initial absorbance value (C_0) by the values measured after a specific incubation time (C).

All experiments were carried out in the dark to avoid photolysis. Temperature was kept constant at 20°C (± 2°C), pH at 7.4 for the duration of the experiments.

Inoculation was achieved with 5-TTri-acclimated activated sludge obtained from stage one of CAS-M in a previous sampling (see 4.2.2). Acclimation was performed over several generations from one to eight in the following manner: First generation reactor was inoculated with 1.0 mL PBS-washed activated sludge and was kept running until biodegradation occurred. The experiment was stopped and 1.0 mL of the reactor used for inoculation of a subsequent experiment referred to as generation two. After observing biodegradation the experiment was stopped and the third generation reactor was prepared in the same manner. These setups were repeated until generation eight, obtaining a biomass highly selected for 5-TTri biodegradation.

Samples for DNA analysis by DGGE were taken after 7, 9 and 22 days by withdrawing 100 mL reactor suspended biomass. Samples for DNA and RNA analyses by next-generation-sequencing were taken at day 9 in the same way. Samples were divided 1:1 and filled into pre-washed glass bottles and prepared differently for DNA and RNA extraction.

6.2.3 DNA and RNA Extraction for Next-Generation-Sequencing

DNA Preparation: 50 mL samples were centrifuged 30 min, 4°C, and 10,000 g. The pellet was resuspended in 500 µL PBS-Buffer (NaCl (8.0 g L^{-1}), KCl (0.2 g L^{-1}), Na$_2$HPO$_4$ (2.7 g L^{-1}), KH$_2$PO$_4$ (0.2 g L^{-1})) and kept frozen at -80°C before extraction.

RNA Preparation: 50 mL samples were centrifuged for 30 min, 4°C, and 10,000 g. The pellet was resuspended with 1000 µL RNAlater RNA Stabilization Reagent (Qiagen). After inoculation for 1 min, 4°C, cells were pelleted again with the same centrifugal settings, RNAlater discarded and the pellet only frozen at -80°C before RNA extraction.

DNA/RNA Extraction: All extractions were performed with all-in-one kits from Qiagen. For DNA the DNeasy Blood & Tissue Kit and for RNA the RNeasy Mini Kit was used. All steps were conducted according to the manufacturer's protocol. DNA and RNA concentrations were measured with a Nanodrop (Thermo Fisher Scientific, Schwerte, Germany). Extracted DNA and RNA were stored in elution buffer at -80°C before further analysis.

6.2.4 Denaturing Gradient Gel Electrophoresis (DGGE)

The V1 – V3 regions of the bacterial 16S rRNA genes were amplified by PCR from the total genomic DNA obtained in the same way as the DNA for next-generation-sequencing. The primers 27f-GC (5'-AGA GTT TGA TCM TGG CTC AG-3') with GC clamp (5'-CGC CCG CCG CGC CCC GCG CCC GTC CGC CGC CGC CCC CCG CCC CGG-3') attached to the 5' end of the primer 517r (5'-

GTA TTA CCG CGG CTG CTG GC -3´) [152]. The V3 – V5 regions were amplified using primers 341f-GC (5´-CCT ACG GGA GGC AGC AG-3´) and 907r (5´-CCG TCA ATT CMT TTG AGT TT-3´). Temperature cycling was performed using GoTaq Hot start master mix from Promega (Mannheim, Germany). One cycle of 95°C for 2 min, 15 cycles of 94°C for 0.5 min, X°C for 0.5 min and 72°C for 0.5 min, followed by 20 cycles of 94°C for 1 min, X°C for 0.5 min and 72°C for 0.6 min (each step elongation time was increased by 5 seconds). X= 55°C (27f/517r); 52°C (341f/907r). DGGE analysis was performed on a DCode universal mutation detection system from Bio-Rad (Munich, Germany). 10 µL of the obtained PCR products were loaded onto 6% (w/v) polyacrylamide gels in 1X TAE buffer using a denaturing gradient ranging from 20% to 80% with a 100% denaturing solution being defined as 7 M urea and 40% (v/v) formamide. Electrophoresis was performed at 60 V for 16 h at 55°C. After electrophoresis, gels were stained for 10 min with ethidium bromide (0.5 µg mL^{-1}) in TAE buffer and photographed. Dominant bands were excised from the gel, eluted in 100 µL sterile TE buffer overnight at 20°C and re-amplified with their corresponding primer set 27f (GC free) and 517r or 341f (GC free) and 907r. The PCR products were cleaned from residual primer with QIAquick PCR Purification Kit (Qiagen) before they were sent for sequencing to MWG Operon (Ebersberg, Germany).

6.2.5 Phylogenetic Analysis

Nucleotide sequences that were obtained from DGGE bands of 16S rRNA genes were analyzed by comparing the gene fragments with the ones listed in the public database ENA (European

Nucleotide Archive, http://www.ebi.ac.uk/ena/) to identify the nearest phylogenetic neighbors. Species names, closest relatives, and maximum similarity are given in Table 6.2.

6.2.6 Library Preparation and Next Generation Sequencing

Metagenomic Analysis by DNAseq

1µg of genomic DNA was used to generate the sequencing libraries following the Truseq DNA v2.0 protocol (Illumina). Genomic DNA was fragmented at a mean size of 300 bp by a Covaris S220. Samples were multiplexed on a single lane of a HiSeq flow cell v3 (Illumina). Paired end reads of 2 x 50 nucleotides were produced with a Hiseq1000 using SBS v3 kits (Illumina). Cluster detection and base calling were performed using RTAv1.13 and quality of reads assessed with CASAVA v1.8.1 (Illumina). The sequencing resulted in 34 and 70 million pairs of 50 nucleotide long reads for each sample with a mean Phred quality score > 35.

Metatranscriptome analysis by RNAseq

Directed RNA-Sequencing libraries were generated from 50ng of total RNA following the Truseq RNA protocol (Illumina, without purification) with polydT beads and the modification of the 2nd strand cDNA synthesis described elsewhere [153]. Paired end reads of 2 x 50 nucleotides were obtained from a single lane with a Hiseq1000 using SBS v3 kits (Illumina). Cluster detection and base calling were performed using RTAv1.13 and quality of reads assessed with CASAVA v1.8.1 (Illumina). The sequencing resulted in 62 and 121 million pairs of 50 nucleotide long reads for each sample with a mean Phred quality score > 35.

6.2.7 Sequence-Data Processing

Andreas Doetsch at the Institute of Functional Interfaces, KIT, Karlsruhe, accomplished all metagenomic and metatranscriptomic data processing.

A *de novo* assembly with the sequence fragments was performed. Therefore, the metagenome of each sample was assembled using idba-ud [154] with k-values ranging from 35 to 50 and step size of 5. Protein coding genes were predicted with MetaGeneMark using default parameters [155] and annotated by BLAST search using BLAST+ version 2.2.28 [156] and the non-redundant protein database (nr) available from NCBI.

The original sequence reads were mapped to the contigs resulting from the *de novo* assembly using bowtie2 [157] and coverage of the predicted protein coding genes was calculated by summing up the number of reads that overlapped with the annotated open reading frames (*rpg* or *reads per gene*). The taxonomic composition of each metagenome was estimated based on the taxonomy information associated with the BLAST-annotation of protein coding genes. The abundance of a certain taxon was determined by summing up all rpg values of genes with identical taxon ID excluding proteins with less than 80 % identity and 80 % alignment length of the best BLAST hit.

6.3 RESULTS AND DISCUSSION

6.3.1 Biodegradation of 5-TTri under Aerobic Conditions and Different Media

5-TTri biodegradation experiments were performed in batch mode, i.e. only initial nutrient supply, and with two generations one and eight. Generation 1 was derived right from activated sludge as no high sensitivity was required, while generation eight was acclimated eight-times to high 5-TTri concentrations (see 5.2.3). The setups containing only MSM without any additional nutrients did not show any biodegradation in both generations and thus were not considered for subsequent analysis (Figure 6.1, ○ and ●). Approaches conducted with MSM-CN and MSM-SS showed different biodegradation patterns (Figure 6.1). 1st and 8th generation setups in MSM-SS showed a steady decrease in UV-absorbance immediately from the start indicating a high 5-TTri biodegradation. First generation in MSM-SS additionally showed an initial slight increase in UV-AM due to biomass growth that was not observed in the 8th generation. This indicates that 8th generation organisms focused on 5-TTri utilization while 1st generation organisms were not well adapted yet and used the sludge supernatant primarily for growth instead of co-metabolically degrading 5-TTri. 1st and 8th generation setups in MSM-CN behaved completely different regarding biodegradation but started off the same way with an increase in UV-AM due to the nutrients in MSM-CN being utilized for biomass production. After three days 8th generation setups showed a strong decrease in UV-AM that indicated 5-TTri biodegradation while parallel 1st generation setup did not show any biodegradation and 5-TTri concentration remained stable.

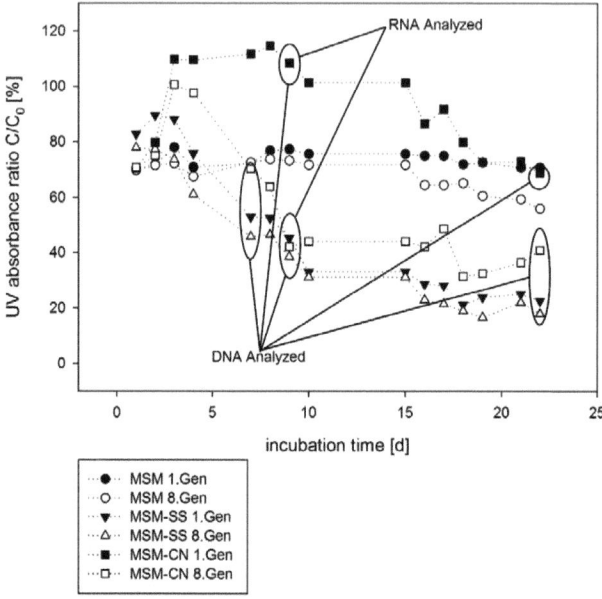

Figure 6.1 – 5-TTri biodegradation patterns observed with UV-AM in three different media (Table 6.1) inoculated with ASC from generation one and eight. DNA analyses with DGGE were performed at day 7, 9 and 22, DNA and RNA analyses with next-generation-sequencing at day 9.

After seven days 1st and 8th generation setups in MSM-SS and 8th generation in MSM-CN showed a strong decrease in UV-AM. Therefore, biomass samples from the seventh day were used for DNA extraction to analyze the ASC composition as it was assumed that the 5-TTri biodegrading ASCs should then have been fully established. Two days later these three setups decreased even further while 1st generation setup in MSM-CN still did not show any sign of biodegradation. Therefore, all four setups were sampled at day nine and were used for DNA as well as RNA extraction because all setups but 1st generation MSM-CN should have been fully active towards 5-TTri biodegradation. These four samples were used for DNA analysis to evaluate the abundance of organisms therein and for RNA analysis to assess the organisms' specific activity. First generation MSM-CN did not show any biodegradation and was thus used as 'blank' to study the ASC composition and activity of a 5-TTri non-degrading bacterial community (Figure 6.1). In addition, DNA samples for DGGE analyses were also taken at day 22 at the end of the experiment.

6.3.2 DGGE Patterns of 5-TTri Biodegrading ASC

The total DNA obtained from the samples at day seven, nine, and at the end of the biodegradation experiment at day 22 were analyzed by DGGE to profile their diversity and detect changes in the ASC structure among the different generations. Two primer sets were used to evaluate if differences in the banding patterns occur due to applied primers.

Primer set 27f_GC-517r revealed a very clear banding pattern that in most samples consisted of no more than three bands (Figure 6.2) that were all located at a gradient concentration of around 70 to 75% (Figure 6.3). MSM-CN 8th generation setups always showed three clear bands with the first band being present in these setups only. Highest diversity was observed in MSM-SS 8th generation setups that showed two clear bands with the first band being a double one and a very strong band at the end of the profile. Many faint bands were detected in between. The total diversity in all setups was very low as all bands were located in a narrow gradient from 70% to 75% whereas not many taxonomic groups are expected. All marked bands (white rectangles, 1 to 3 in Figure 6.2) were excised from the gel and taxonomically identified (Table 6.2).

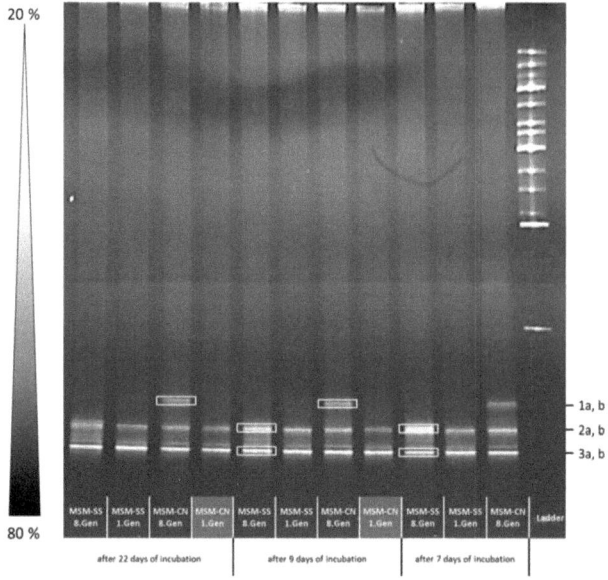

Figure 6.2 – DGGE band profiles of the ASC from generation one and eight after seven, nine and 22 days of incubation (see Figure 6.1) in two different media (Table 6.1).

PCR products, amplified with primers 27f_GC-517r, were analyzed on a gradient ranging from 20 to 80%. A 1kb DNA ladder was used as marker. Marked bands were excised and sequenced (see Table 6.2).

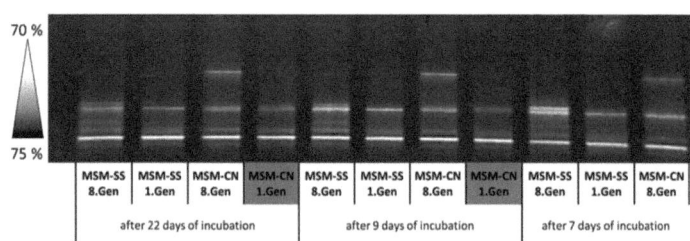

Figure 6.3 – Magnification of DGGE band profiles of ASC amplified with primer set 27f_GC-517r. Visible gradient ranging from 70 to 75%.

Table 6.2 – Species found in the 5-TTri biodegrading ASC by sequence analysis of excised DGGE bands. Provided are sequence length and the similarity to the closest relative in the ENA database (http://www.ebi.ac.uk/ena/).

Primer/ Figure	DGGE band	generation	Species	similarity [%]	closest relative (ENA)
27f_GC – 517r (Figure 6.2)	1a	8	Pseudomonas sp.	100	KF379759, Pseudomonas fragi
	1b	8	Pseudomonas sp.	100	KF379759, Pseudomonas fragi
	2a	8	Flavobacterium sp.	99	JQ349044, Flavobacterium sp. KTce-4
	2b	8	Flavobacterium sp.	99	JQ349044, Flavobacterium sp. KTce-4
	3a	8	Hydrogenophaga sp.	99	AB638427, Hydrogenophaga defluvii
	3b	8	Hydrogenophaga sp.	98	AB638427, Hydrogenophaga defluvii
341f_GC – 907r (Figure 6.4)	4a	8	Pseudomonas sp.	99	AY972176, Pseudomonas plecoglossicida
	4b	8	Pseudomonas sp.	99	AY972176, Pseudomonas plecoglossicida
	4c	8	Pseudomonas sp.	99	AY972176, Pseudomonas plecoglossicida
	4d	8	Pseudomonas sp.	99	AY972176, Pseudomonas plecoglossicida
	5a	1	Aminobacter sp.	99	NR025301, Aminobacter aminovorans
	5b	1	Aminobacter sp.	99	NR025301, Aminobacter aminovorans
	6a	1	Hydrogenophaga sp.	99	AB638427, Hydrogenophaga defluvii
	6b	8	Hydrogenophaga sp.	99	AB638427, Hydrogenophaga defluvii
	7a	1	Aminobacter sp.	99	KC767647, Aminobacter aminovorans
	7b	1	Aminobacter sp.	99	KC767647, Aminobacter aminovorans

They revealed two *Pseudomonas* sp. (1 a, b), two *Flavobacterium* sp. (2 a, b) and two *Hydrogenophaga* sp. (3 a, b). The parallel bands a and b always revealed the same organism showing that

DGGE separation was working properly. Again, different *Pseudomonas* spp. were identified. They are important for the biodegradation of various organic micropollutants including sulfamethoxazole, chlorinated compounds and several complex dyes [158-160]. Thus, this bacterial group might also play a crucial role in 5-TTri biodegradation. Also the other two organisms have the ability to biodegrade complex organic compounds as was shown in literature for 4-aminobenzenesulfonate biodegradation with *Hydrogenophaga* sp. [161] and 1,4-dioxane with a *Flavobacterium* sp. [162]. In this study, it was not possible to cultivate *Pseudomonas* sp. and *Flavobacterium* sp. in pure culture while *Hydrogenophaga* sp. could be isolated (data not shown). Thus, specific biodegradation experiments with a consortium containing all three bacteria could not be performed. As biodegradation experiments with *Hydrogenophaga* sp. pure cultures did not show any biodegradation (data not shown) it seems very likely that these three species play an important role for 5-TTri biodegradation by using concerted metabolic processes or by formation of concerted syntrophic working environment with one species relying on the other during 5-TTri removal.

A closer look at the DGGE bands showed distinct changes in the banding pattern over the two generations analyzed (Figure 6.3). 8th generation setups with MSM-SS showed a shift in the upper double band. This double band was identified as *Flavobacterium* sp.. After seven days of incubation, the double band was very sharp and intense. After nine days the uppermost double band already fainted out and was hardly visible while after 22 days of incubation these double bands were almost completely gone and visible as faint shadow only. The lowermost single band, representing *Hydrogenophaga* sp. and detected in all setups, did not show any change in the different setups and during cultivation. Therefore, *Hydrogenophaga* sp. might be crucially necessary for 5-TTri biodegradation whereas it might be less important which other organism is present to establish a stable biodegrading community. *Hydrogenophaga* sp. alone was not able to biodegrade 5-TTri as observed with pure cultures (data not shown), explaining why its band occurred in the 5-TTri non-biodegrading setups MSM-CN 1st generation. The first generation MSM-CN setup was obviously not able to establish a stable community and therefore did not show any biodegradation. If one necessary species is missing, the whole community can achieve no biodegradation.

In addition, primer set 341f_GC-907r was applied to evaluate the reliability of the DGGE method and to detect differences in banding patterns due to the primers' amplifying a different region. All bands were located at a gradient ranging from 60 to 65% (Figure 6.4). Again, no more than three clear bands were observed in all setups indicating a rather low biodiversity in the ASC. A very sharp double band was observed for biodegrading 8th generation setups in MSM-CN while non-degrading 1st generation setups completely lacked these bands. These double bands (4 a to d, Figure 6.4) were identified as *Pseudomonas* sp., a behavior known from literature [163].

Therefore, as *Pseudomonas* sp. was only found in biodegrading setups (Figure 6.2, Figure 6.4), they might be participating in 5-TTri biodegradation. Also 1st and 8th generation setups in MSM-SS showed clear differences. First generation MSM-SS showed a sharp band at the same position as non-degrading 1st generation in MSM-CN labeled 5 a and b (Figure 6.4) that was not detected in the 8th generation MSM-SS.

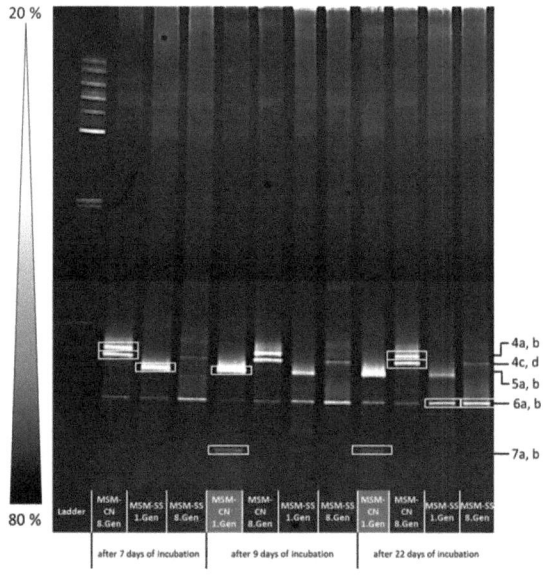

Figure 6.4 – DGGE band profiles of the ASC, amplified with primer set 341f_GC-907r, from generation one and eight after seven, nine and 22 days of incubation and two different media (Table 6.1).

PCR products were and analyzed on a gradient ranging from 20 to 80%. A 1kb DNA ladder was used as marker. Marked bands were excised and sequenced (Table 6.2).

In contrast, 8th generation MSM-SS showed a faint but sharp band at the same position as the lower double band observed for 8th generation in MSM-CN labeled 4 c and d. While bands 4 c and d were identified as *Pseudomonas* sp., bands 5 a and b (Figure 6.4) were identified as *Aminobacter* sp., a species that was not found with the first primer set 27f-517r but is also known from literature for its ability to degrade 2,6-dichlorobenzamide [164]. All setups showed a more or less sharp band at around 65% (6 a and b, Figure 6.4) indicating that this organism is omnipresent in all setups and might be necessary for 5-TTri biodegradation. Sequence analysis revealed a *Hydrogenophaga* sp. as already found with the first primer set verifying that species' importance for 5-TTri biodegradation. The lowermost two bands (7 a and b in contrast to 5 a and b) were also identified as *Aminobacter* sp. (Table 2). This could mean that two different *Aminobacter* spp. were present or the DNA sequence damaged as the lower band is far more faint [165].

These results already indicate the dominant species that maybe contribute to 5-TTri biodegradation. However, a known problem in DGGE is that it detects also dead or inactive cells that might still be present in the setup, as they still contain DNA. This can result in false positive signals, i.e. these cells still produce a band, as described in literature [166]. Furthermore, DGGE mainly detects the dominant organisms present in the setups, as rare species are mostly outcompeted during PCR [167, 168]. In addition, PCR-based studies are inherently biased as not all rRNA genes can be amplified with the same "universal" primers [169]. However, these rare species could contribute to 5-TTri biodegradation but cannot be detected with DGGE as they produce no or no clear signal. Therefore, techniques with a higher sensitivity like next-generation-sequencing analyses are a vital approach to detect also scarce organisms.

6.3.3 Metagenomic Analysis of 5-TTri Biodegrading ASC

This approach revealed the total DNA abundance of the 5-TTri biodegradation setups to evaluate the amount of possible species and their dominance structure in the ASC. The *de novo* sequence alignment was performed against the complete NCBI database. The 10 most abundant genera, regarding their DNA and RNA abundance, are displayed in Figure 6.5 separated in first and 8th generation setups and the two used media MSM-SS and MSM-CN at day nine. 'All the rest' are groups that occurred less abundant.

The most dominant species in 1st generation MSM-CN setups were *Hydrogenophaga* spp. (3.3%), *Mesorhizobium* spp. (44.9%) and *Acidovorax* spp. (0.8%). In MSM-SS 1st generation they also represented the most dominant species but with a lower abundance of *Mesorhizobium* spp. (11.0%) and a higher abundance of *Hydrogenophaga* spp. (8.0%) and *Acidovorax* spp. (1.7%). As 1st generation in MSM-CN showed no biodegradation, it might be that *Mesorhizobium* spp. is not participating in 5-TTri biodegradation but only using the supplied nutrients for growth. This becomes clear when comparing the 1st generation MSM-CN/MSM-SS setups to the 8th generation. In both media, *Mesorhizobium* spp. is not present any longer while *Hydrogenophaga* spp., *Pseudomonas* spp. and *Acidovorax* spp. are by far the most abundant species in the eighth generation MSM-CN/MSM-SS setups. This finding perfectly fits the DGGE results and indicates again that *Pseudomonas* spp. might be crucial for 5-TTri biodegradation but also shows that *Pseudomonas* spp. is efficiently utilizing the specific nutrients acetate and ammonium nitrate as indicated by a higher abundance in MSM-CN (29.1%) than MSM-SS (7.5%).

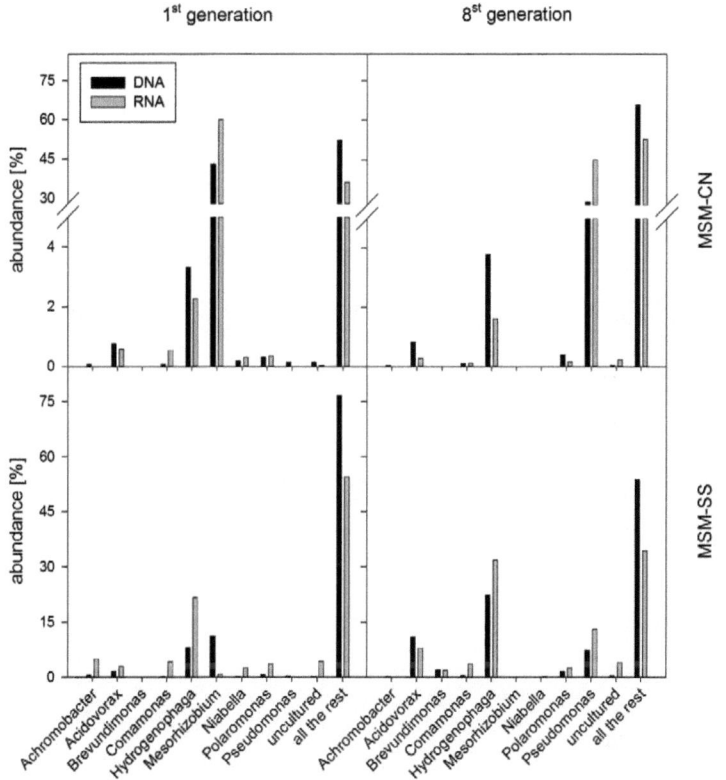

Figure 6.5 – DNA and RNA analyses of the *de novo* assembly of the sequences mapped against the complete NCBI database.
First and 8th generation setups in MSM-SS and MSM-CN were analyzed only on day nine of the biodegradation experiment. Shown are the 10 most abundant genera of generation one. 'All the rest' are groups that occurred less abundant.

In MSM-SS, the distribution of the species is more homogeneous as the complex nutrients in MSM-SS obviously support the growth and activity of various bacteria. This explains why 5-TTri biodegradation in MSM-SS occurred faster compared to MSM-CN. It might be that 5-TTri is biodegraded by several species working syntrophically together. Probably many different species are able to utilize 5-TTri when the right organisms accompany them. In addition, a principal component analysis of the ASC in the different setups showed an even clearer distribution (Figure 6.6 – Principal component analysis of the DNA (○) and RNA (●) extracted from the 5-TTri biodegrading setups at day nine.). The non-biodegradation group (Figure 6.6, left side) consisted mainly of *Mesorhizobium* spp. while the dominant organism in the MSM-CN biodegradation group was identified as *Pseudomonas* spp. (Figure 6.6, right side). This again indicates the importance of *Pseudomonas* spp. for 5-TTri biodegradation as it appears that *Pseudomonas* spp. have switched places with *Mesorhizobium* spp. from generation one to eight. *Mesorhizobium* spp., the most abundant species from non-biodegrading 1st generation MSM-CN were hardly found any more in the biodegrading 8th generation MSM-CN while *Pseudomonas* spp. behaved the opposite: Hardly present in MSM-CN 1st generation, *Pseudomonas* spp. showed a high abundance in MSM-CN 8th generation. *Mesorhizobium* spp., a common organism in activated sludge [170], might be present in a high abundance as 1st generation ASC was derived right from the WWTPs' activated sludge, but might not play a role for 5-TTri biodegradation. Fastest biodegradation occurred in the MSM-SS-supplied group that is clearly separated from the other two groups (Figure 6.6, bottom). The most dominant organisms therein were *Hydrogenophaga* spp. and *Acidovorax* spp.. MSM-SS setups did not show such a shift in abundance over the two generations. This shows that MSM-SS 1st

generation setups already established a fully functional 5-TTri biodegrading community without the explicit need for further improvements. MSM-SS might contain nutrients that could induce a higher metabolic activity in the ASC rather than fostering growth of specific organisms as observed in MSM-CN.

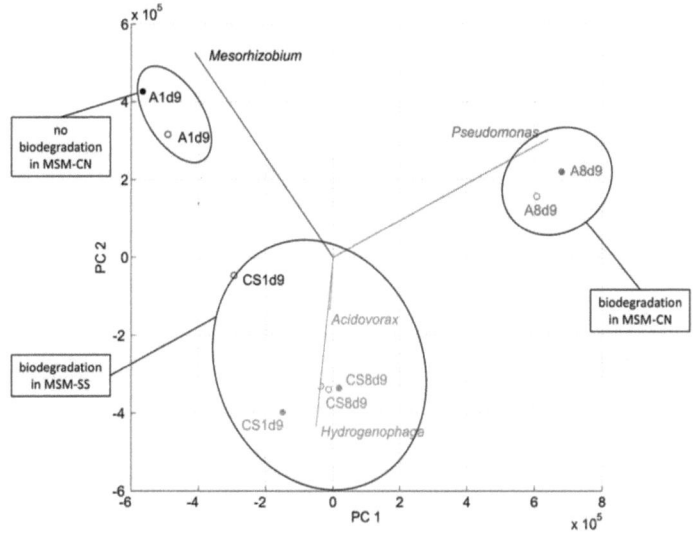

Figure 6.6 – Principal component analysis of the DNA (○) and RNA (●) extracted from the 5-TTri biodegrading setups at day nine.
A= MSM-CN; CS= MSM-SS; XdX= generation, day X (A1d9= MSM-CN, first generation, sample from day nine)

These results, compared to the organisms detected with DGGE, show that the total diversity of the ASCs was still high but many organisms were present in a low abundance only. Maybe, effective 5-TTri biodegradation is depending on the establishment of a biodegrading community. To conclude, the most likely organisms contributing to biodegradation according their DNA abundance are

Acidovorax spp., *Hydrogenophaga* spp. and *Pseudomonas* spp.. However, to correlate their DNA abundance with their RNA abundance, i.e. their activity, metatranscriptomic analyses have to follow.

6.3.4 Metatranscriptomic Analysis of 5-TTri Biodegrading ASC

Metagenomic data reveals the organisms' DNA abundance and therefore the total ASCs' diversity while metatranscriptomic analysis reveals the organisms' RNA abundance and allows a correlation with their specific activity. The most abundant organisms are not necessarily the most active ones as also dead or inactive cells are detected with metagenomic analyses. Such cells, even if they show a high DNA abundance, observe a low RNA abundance, i.e. a low activity, as seen in the 'all the rest' group (Figure 6.5). This group represents more than 50% of the DNA abundance and contains >1,000 different species but shows always a significant lower RNA abundance and thus activity.

Mesorhizobium spp. were the most abundant organisms in 1st generation MSM-CN and their activity, observed with RNA analysis (Figure 6.5), was even higher. No biodegradation was found in that setup indicating that their activity might be due to general metabolic processes not related to 5-TTri biodegradation. In contrast, *Hydrogenophaga* spp. showed a lower specific RNA activity compared to its DNA abundance indicating only general metabolism on the nutrients but no 5-TTri biodegradation activity. Compared to first generation MSM-SS setup, which already showed 5-TTri biodegradation potential, *Hydrogenophaga* spp.

showed a 10-fold higher activity while the activity of *Mesorhizobium* spp. decreased by 86-fold.

Eighth generation setups showed a similar pattern regarding *Hydrogenophaga* spp. with a decreased activity in MSM-CN and a significantly increased activity in MSM-SS. Another dominant genus, only occurring in eighth generation setups in both media, was *Pseudomonas* spp.. This organism showed a very high activity in MSM-CN (44.8%) and a reduced in MSM-SS (13.1%). It might be that *Pseudomonas* spp. do not biodegrade 5-TTri alone and the formation of a biodegrading ASC is required. As *Pseudomonas* spp. exhibited intensive growth and presumably consumed most of the nutrients supplied in MSM-CN, the formation of a biodegrading community might be aggravated explaining the delay in biodegradation in eighth generation MSM-CN setup (Figure 6.1). The third dominant genus in eighth generation setups, i.e. *Acidovorax* spp., also observed a higher activity in MSM-SS compared to MSM-CN indicating that complex nutrient MSM-SS fosters metabolic activity. This might allow the organisms easily establishing a functioning biodegrading ASC as complex MSM-SS nutrients enhance growth and metabolism of a larger variety of organisms.

In general, when comparing DGGE and next-generation sequencing results, it was observed that DGGE was limited in its resolution and identification of organisms. DGGE analyses revealed around four organisms and no RNA analyses were possible. Therefore, DGGE results did not reveal organisms' dominance structures or activities. In contrast, next-generation sequencing (NGS) provided a fast and sensitive technique to screen and detect the most abundant and most active species in ASCs. Analyses showed a high diversity and allowed a closer look at species level. However, DGGE and NGS showed comparable results as the dominant organisms were detected with both, DGGE and NGS. Thus, DGGE is still a powerful tool to access the dominant species of ASCs but NGS might be the method of choice when information about RNA activity is needed. Nevertheless, none of the techniques revealed the ASC composition necessary for 5-TTri biodegradation. To specifically identify the biodegrading organism, pure cultures are inevitable as shown for sulfamethoxazole biodegradation in the following chapter. However, even intensive efforts to gain pure cultures did not reveal the biodegrading organism indication that a biodegrading community might be more likely for 5-TTri biodegradation.

6.4   CONCLUSIONS

Two activated sludge community generations (one and eight), capable of biodegrading 5-tolyltriazole (5-TTri), were characterized by two different methods: denaturing gradient gel electrophoresis (DGGE) and next-generation-sequencing (NGS). DGGE showed a low diversity in the ASC as only the most dominant organisms were detected. Application of two different primer sets revealed different banding patterns but a subsequent sequence analysis of the dominant bands revealed the same organisms except two genera, i.e. *Aminobacter* spp. and *Flavobacterium* spp.. *Hydrogenophaga* spp. and *Pseudomonas* spp. occurred in all setups indication their importance for 5-TTri biodegradation. In addition, metagenomic and metatranscriptomic analyses were performed. They revealed a high biodiversity and further strengthened *Pseudomonas* spp. to be crucial for 5-TTri biodegradation. Comparing species DNA abundance with their RNA activity revealed *Hydrogenophaga* spp., *Acidovorax* spp. and *Pseudomonas* spp. to be important for 5-TTri removal. These results indicate that 5-TTri might not be biodegraded by just one species but rather by a stable ASC consisting of several species performing a concerted syntrophic metabolic process.

Chapter 7

Taxonomic Identification of Pure Cultures from Sulfamethoxazole-Acclimated Activated Sludge with Respect to Sulfamethoxazole Biodegradation Potential

Sulfamethoxazole biodegradation by activated sludge (AS) communities is only partly understood. This chapter is focusing on nine different bacteria species capable of SMX biodegradation isolated from SMX-acclimated AS communities. Pure cultures, isolated from activated sludge, were screened by UV-absorbance measurements (UV-AM) for their SMX biodegradation potential. Identification via almost complete 16S rRNA gene sequencing revealed five Pseudomonas spp., one Brevundimonas sp., one Variovorax sp. and two Microbacterium spp.. Thus, seven species belonged to the phylum Proteobacteria and two to Actinobacteria. These cultures were subsequently incubated in media containing 10 mg L^{-1} SMX and different concentrations of carbon (sodium-acetate) and nitrogen (ammonium-nitrate). Different biodegradation patterns were revealed with respect to media composition and bacterial species. Biodegradation, validated by LC-UV measurements to verify UV-AM, occurred very fast with 2.5 mg L^{-1} d^{-1} SMX being biodegraded in all pure cultures. Media containing only SMX as carbon and nitrogen source proved the organisms' ability to use SMX as sole nutrient source.

7.1 INTRODUCTION

Widespread usage, disposal around the world and a consumption up to 200,000 t per year, makes the various groups of antibiotics an important issue for micropollutant risk assessment [171, 172] and has become of major concern for environmental protection strategies. Antibiotics are designed to inhibit microorganisms and influence microbial communities in different ecosystems [173, 174]. Monitoring programs already showed that antibiotics can be found almost everywhere in the environment in concentrations up to µg L^{-1} [19, 175-178]. Furthermore, antibiotic resistance genes might be transferred to human-pathogenic organisms by horizontal gene-transfer and become a serious issue, especially multidrug resistance in bacteria [16, 179, 180]. Sulfamethoxazole (SMX) is one of the most often applied antibiotics [181]. The frequent use of SMX results in wastewater concentrations up to µg L^{-1} and surface water concentrations in the ng L^{-1} scale [182-185]. Even in groundwater SMX was found at concentrations up to 410 ng L^{-1} [184]. These SMX concentrations might be too low for inhibitory effects as the MIC$_{90}$ for *M. tuberculosis* was found to be 9.5 mg L^{-1} [186], but they might be high enough to function as signalling molecule to trigger other processes like quorum sensing in environmental microbial communities [187]. Different studies showed [188-191] SMX to induce microbial resistances and reduces microbial activity/diversity arising the need for a better understanding of SMX biodegradation. SMX inflow concentrations in WWTPs in µg L^{-1} combined with often partly elimination ranging from 0% to 90% [174, 176, 183, 192] result in high effluent discharge into the environment. To extent knowledge about removal regarding biodegrading microorganisms is implicitly required to optimize environmental nutrient conditions for SMX removal and degradation rates. It is known that SMX can be removed by photodegradation [193, 194] and sorption processes in activated sludge systems [195]. However, biodegradation is, especially in WWTPs, probably the major removal process. Literature data focusing on SMX biodegradation in lab scale experiments with activated sludge communities (ASCs) and pure cultures showed a high fluctuation from almost complete SMX elimination [50, 134, 178] to hardly any removal of SMX [196]. The determined SMX biodegradation potential was clearly affected by nutrient supply whereas this study's emphasis is on clarifying the effect of addition of readily degradable carbon and/or nitrogen sources significantly enhanced SMX elimination [197] while in others supplementation showed no effect. For this purpose pure cultures were isolated from SMX-acclimated ASCs and identified in respect to taxonomy and biodegradation capacity. Aerobic SMX biodegradation experiments with different species were carried out at various nutrient conditions to screen biodegradation potential and behaviour as a base for future research on biodegradation pathways.

7.2 MATERIALS AND METHODS

7.2.1 Chemicals and Glassware

Sulfamethoxazole (SMX, 99.8 % purity) was purchased from Sigma Aldrich (Steinheim, Germany), all other organic media components were from Merck KGaA (Darmstadt, Germany) while the inorganic media components were purchased from VWR (Darmstadt, Germany). High-purity water was prepared by a Milli-Q system (Millipore, Billerica, MA, USA). All glassware used was procured from Schott AG (Mainz, Germany) and pre-cleaned by an alkaline detergent (neodisher®, VWR Darmstadt, Germany) followed by autoclaving for 20 min at 121°C.

7.2.2 Activated Sludge Sampling

Activated sludge was taken as grab sample from stage 1 of a 2-stage municipal conventional activated sludge plant (CAS-M), located near the city of Munich, Germany and treating 1 million population equivalents. Stage 1 is the high load stage with a food to microorganism ratio of 0.64 kg BOD_5 kg^{-1} $MLSS^{-1}$. The influent consists of municipal and industrial wastewater (1:1). 500 mL AS were collected in pre-cleaned 1L glass bottles, stored at 4°C and used within 24h for inoculation of the different setups.

7.2.3 Experimental Setup

7.2.3.1 SMX-acclimated ASCs

Evaluation of AS biodegradation potential obtained from the WWTP, was performed in 150 mL R2A-UV media (casein peptone 1,000 mg L^{-1}, glucose 500 mg L^{-1}, potassium phosphate 300 mg L^{-1}, soluble starch 300 mg L^{-1}, DOC:N ratio 7:1, pH 7.4), spiked with 10 mg L^{-1} SMX to apply a high selective pressure. Non-SMX-resistant organisms were ruled out and the chance to obtain SMX biodegrading organisms was increased in subsequent isolation steps. After biodegradation occurred the experiment was stopped and the remaining biomass was used to inoculate a second setup under the same conditions to further decrease microbial diversity and favor SMX-resistant/biodegrading organisms. After the second setup showed biodegradation, the experiment was stopped and the biomass used for cultivation of SMX biodegrading organisms on solid R2A-UV media (1.5% agar supply). SMX removal was determined by UV-absorbance measurements (UV-AM) as fast pre-screening method for biodegradation (see 2.4.1).

7.2.3.2 Cultivation and Isolation of Pure Cultures

Pure cultures were successfully cultivated and isolated from SMX-acclimated biodegrading ASCs. 200 µL AS was plated on solid R2A-UV media containing 10 mg L^{-1} SMX to inhibit growth of non-resistant bacteria and foster growth of potential SMX-resistant/biodegrading organisms. After cultures were observed on solid media they were isolated and further purified by streaking on new plates resulting in 110 isolates. These were used for inoculation of 100 mL setups with 20 mL MSM-CN media (KH_2PO_4 80 mg L^{-1}, K_2HPO_4 200 mg L^{-1}, Na_2HPO_4 300 mg L^{-1}, $MgSO_4$*7 H_2O 20 mg L^{-1}, $CaCl_2$*2 H_2O 40 mg L^{-1}, $FeCl_3$*6 H_2O 0.3 mg L^{-1}, sodium acetate 300 mg L^{-1} and NH_4NO_3 7.5 mg L^{-1}, DOC:N ratio 33:1, pH 7.4) spiked with 10 mg L^{-1} SMX. Setups were monitored with UV-AM (see 2.4.1) for possible biodegradation. Isolates showing biodegradation were further identified by 16S rRNA gene sequence analysis.

7.2.3.3 Biodegradation Setups with Pure Cultures

Batch experiments were performed to A) screen for biodegradation potential in the isolated cultures and B) determine differences in SMX biodegradation pattern and rate concerning the availability of nutrients. Three media, R2A-UV, MSM-CN and MSM (as MSM-CN but without sodium acetate and NH_4NO_3) were used and inoculated with pure cultures in 100 mL setups filled with 20 mL of media spiked with 10 mg L^{-1} SMX. Duplicate setups (n=2) including sterile, i.e. autoclaved biomass and abiotic, i.e. without biomass, controls for each medium were prepared. Aerobic conditions and photolysis prevention were ensured by shaking at 150 rpm on an orbital shaker in the dark.

The setups were sampled once a day for MSM-CN and MSM media and twice a day for R2A-UV, by taking 1 mL supernatant after half an hour of sedimentation that was sufficient to ensure not to withdraw much biomass. 200 µL was used for UV-AM and 800 µL for LC-UV measurements.

7.2.4 Analyses of Sulfamethoxazole

7.2.4.1 UV-AM

200 µL were taken from the setups and directly used for UV-AM at a wavelength of 257 nm (chapter 2) with the following changes applied. Calibration was performed with 1.0, 5.0, 10.0 and 15.0 mg L^{-1} SMX in high-purity water and the used media. 96 well UV-star plates from Greiner (Greiner Bio-One GmbH, Frickenhausen, Germany) filled with 200 µL were used for measurements and analyzed with an automated plate reader (EnSpire® Multimode

Plate Reader, Perkin Elmer, Rodgau, Germany). Each measurement included an SMX blank (media with SMX but without organisms) was measured to detect changes over time as well as a blank (media without SMX) to detect background absorbance.

7.2.4.2 LC-UV Analysis

800 µL samples obtained from the setups were centrifuged (10 min, 8,000 g, 20°C), filtrated through a 0.45 µm membrane filter to remove cellular debris and biomass and filled into sterile glass flasks. Flasks were stored in the dark at -20°C before analysis.

A Dionex 3000 series HPLC system (Dionex, Idstein, Germany), equipped with an autosampler, was used to perform a DAD scanning from 200 to 600 nm to detect and quantify SMX. Chromatographic separation was performed on a Nucleosil 120 - 3 C18 column (250 mm x 3.0 mm i.d., 3 µm particle size) from Macherey Nagel (Düren, Germany). Column temperature was 25 °C. The mobile phases were acetonitrile (AN) and water (pH 2.5 using phosphoric acid). The gradient used was 0-5 min, 7% AN; 5-18 min, 7-30% AN; 18-30 min, 30% AN; 30-35 min, 7% AN. The solvent flow rate was 0.6 mL min^{-1}. The column was allowed to equilibrate for 5 min between injections. Limit of quantification and limit of detection were 0.1 mg L^{-1} and 0.03 mg L^{-1}, respectively.

7.2.5 Taxonomic and Phylogenetic Identification of Isolated Pure Cultures by 16S rRNA Gene Sequence Analysis

DNA of SMX biodegrading organisms was extracted by a standard phenol/chloroform/CTAB extraction method. 16S rRNA gene was subsequently amplified via standard PCR using universal bacterial primers 27f (5-AGA GTT TGA TCM TGG CTC AG-3) and 1492r (5-TAC GGY TAC CTT GTT ACG ACT T-3) [198]. All cultures were sent to MWG Operon (Ebersberg, Germany) for sequencing using again primers 27f and 1492r, resulting in nearly full length 16S rRNA gene sequences. Sequences were analyzed with and submitted to the European Nucleotide Archive (http://www.ebi.ac.uk/ena/) (Table 7.2). Subsequent phylogenetic analysis was accomplished with the sequences using the alignment and tree calculation methods of the ARB software package [199]. The nearly complete 16S rRNA gene sequences of the species isolated in this study and their corresponding published closest relatives (http://blast.ncbi.nlm.nih.gov/Blast.cgi) were added to an existing ARB-alignment for the 16S rRNA gene sequence. Alignment was performed with the CLUSTAL W implemented in ARB. Phylogenetic trees of the 16S rRNA gene sequences were calculated based on maximum likelihood.

7.3 RESULTS

7.3.1 SMX Biodegradation

7.3.1.1 Cultivation and Evaluation of Pure Cultures Biodegradation Potential

Isolation of pure cultures was accomplished from SMX-acclimated ASC as described in chapter 2.2.3.1. Growth of cultures on solid R2A-UV media, spiked with 10 mg L^{-1} SMX, was controlled every 24 hours. All morphologically different colonies were streaked onto fresh R2A-UV agar plates, finally resulting in 110 pure cultures. For identification of potential SMX biodegrading cultures, all 110 isolates were inoculated into 20 mL MSM-CN media. SMX biodegradation, shown in Figure 7.1 for 30 isolates as an example, was controlled every two days. After two days a decrease in absorbance was already detected in five cultures followed by seven more at day 4 and 6 while the remaining cultures showed no change. The experiment was stopped after 21 days (just 10 days shown, Figure 7.1), revealing no further SMX biodegrading culture. A 50 % cutoff line (Figure 7.1) was defined as a decrease in UV-absorbance being significant enough to ensure biodegradation. 12 organisms showed a decrease in absorbance greater than 50 % of the initial value. They were further taxonomically identified and used for subsequent biodegradation experiments.

Table 7.1 – Initial (10 mg L^{-1}) and end concentrations of SMX accomplished with 12 biodegrading pure cultures gained out of 110 cultures. Taxonomic identification succeeded with BLAST (http://blast.ncbi.nlm.nih.gov/Blast.cgi). *duplicate organisms; all but SMX344 were discarded.

pure culture	SMX conc. after 10 days [mg L^{-1}]	pure culture	SMX conc. after 10 days [mg L^{-1}]
Brevundimonas sp. SMXB12	0.00	Pseudomonas sp. SMX 333*	1.09
Microbacterium sp. SMXB24	0.00	Pseudomonas sp. SMX 336*	4.35
Microbacterium sp. SMX348	0.00	Pseudomonas sp. SMX 342*	1.09
Pseudomonas sp. SMX321	0.68	Pseudomonas sp. SMX344*	0.23
Pseudomonas sp. SMX330	0.68	Pseudomonas sp. SMX345	1.58
Pseudomonas sp. SMX331	2.68	Variovorax sp. SMX332	3.53

Additionally, biodegradation of these 12 identified isolates was validated by LC-UV (Table 7.1). For cost efficiency only initial- and end-concentrations of SMX in the media were determined as absorbance values did not change any more (Figure 7.1). A decrease in SMX concentration from initially 10 mg L^{-1} to below 5 mg L^{-1} was detected for all 12 isolates (Table 7.1) after 10 days of incubation. It was demonstrated that only 3 cultures eliminated all 10 mg L^{-1} SMX completely while the residual SMX concentrations for the remaining cultures ranged from 0.23 to 4.35 mg L^{-1} after 10 days of incubation.

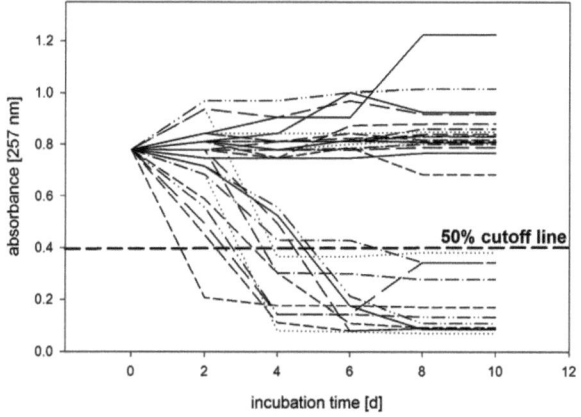

Figure 7.1 – Absorbance patterns of isolated pure cultures in MSM-CN.
Shown are 30 out of 110 isolated pure cultures including the 12 cultures showing SMX biodegradation potential. Initial SMX concentration was 10 mg L^{-1}. The 50 % cutoff line indicates a possible 50 % SMX biodegradation. Each dotted line represents one isolate. All cultures showing more than 50 % decrease in absorbance were further identified.

7.3.2 Taxonomic and Phylogenetic Identification of Pure Cultures

All 12 cultures were identified by 16S rRNA gene sequence analysis to evaluate their phylogenetic position and closest relative. Four cultures, SMX 332, 333, 336 and 344, turned out to be the same organism closely related to *Pseudomonas* sp. He

(AY663434) with a sequence similarity of 99%. Only SMX 344 was kept for further experiments as it showed fastest biodegradation in pre-tests (Figure 7.1 and Table 7.1). Hence, a total of 9 different bacterial species with SMX biodegradation capacity were obtained. Their accession numbers, genus names and closest relatives as found in the NCBI database (http://blast.ncbi.nlm.nih.gov/Blast.cgi), are shown as a maximum likelihood-based phylogenetic tree (Figure 7.2) evaluated with 16S rRNA gene sequence comparisons [200].

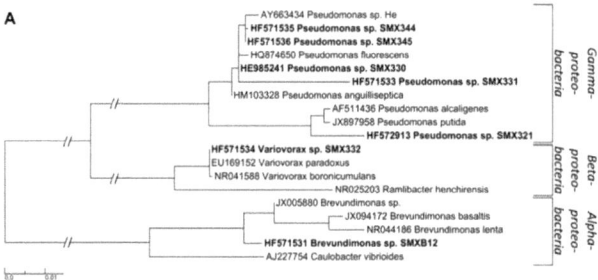

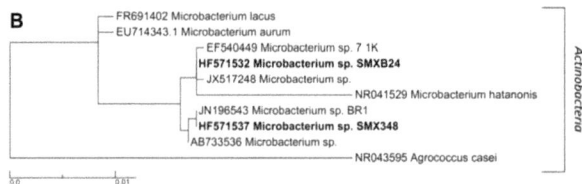

Figure 7.2 A and B – Maximum likelihood-based trees reflecting the phylogeny and diversity of the isolated nine species capable of SMX biodegradation.
Trees are based on nearly complete 16S rRNA gene sequence comparisons. Phylogenetic tree calculated for A) *Pseudomonas* spp., *Variovorax* spp. and *Brevundimonas* spp. and B) for *Microbacterium* spp.. The tree shows the sequences obtained in this study (bold text) and their next published relatives according to the NCBI database (plain text). Numbers preceding taxonomic names represent EMBL sequence accession numbers. Scale bar indicates 0.01 % estimated sequence divergence.

Seven of the nine isolates are affiliated within the phylum *Proteobacteria* represented by the classes *Alpha-*, *Beta-* and *Gammaproteobacteria*, while two belonged to the Phylum *Actinobacteria*. The phylogenetic positions of seven isolated pure cultures affiliated within the phylum Proteobacteria were located in the same tree (Figure 7.2 A). Five different *Pseudomonas* spp. were identified and form two seperate clades representing a highly diverse group. *Pseudomonas* sp. SMX344 and 345 is building an individual cluster but belonged to the same group as SMX330 and 331. All four are closely related to *P. fluorescens* but SMX331 showed a remarkable difference. In contrast to the described *Pseudomonas* spp. above, *Pseudomonas* sp. SMX321 clusters together with *P. putida* and *P. alcaligenes* but forms an individual branch. The other two *Proteobacteria* identified pure cultures belonged to the genera *Variovorax* (SMX332) and *Brevundimonas* (SMXB12). The isolated *Variovorax* SMX332 fell into the *Variovorax paradoxus/boronicumulans* group with a sequence similarity >99% to *V. paradoxus* (EU169152).The *Brevundimonas* sp. SMXB12 was clearly separated from its closest relatives *Brevundimonas basaltis* and *B. lenta* and formed its own branch.

Both *Actinobacteria* affiliated pure cultures were identified as *Microbacterium* spp. and were embedded in a new phylogenetic tree as their phylogenetic position was too far from the other isolates (Figure 7.2 B). The two isolated species were affiliated to two different clades clearly separated from *M. lacus* and *M. aurum*. *Microbacterium* sp. SMXB24 fell into the same group as *Microbacterium* sp. 7 1K and *M. hatatonis* but the branch length clearly showed separation. *Microbacterium* sp. SMX348 was closely related with a sequence similarity of >99% to *Microbacterium* sp. BR1 which was found to biodegrade SMX in an acclimated membrane bioreactor [104].

7.3.3 SMX Biodegradation Studies with Pure Cultures

Setups with sterile biomass (autoclaved) and without biomass (abiotic control) proved SMX to be stable under the operating conditions.

Table 7.2 – Biodegradation rates of the cultures able to biodegrade SMX. Isolation was performed from an SMX-acclimated ASC, followed by identification with 16S rRNA sequencing. ENA accession numbers and species names are provided. * calculated from duplicate experiments (n=2). Standard deviations between duplicate setups were below 1% and are not shown.

accession/isolate	phylum	biodegradation rates* [mg L^{-1} d^{-1}]		
		R2A-UV	MSM-CN	MSM
HF571531, *Brevundimonas* sp. SMXB12	*Proteobacteria*	2.5	1.7	1.0
HF571532, *Microbacterium* sp. SMXB24	*Actinobacteria*	2.5	1.25	1.25
HF571537, *Microbacterium* sp. SMX348	*Actinobacteria*	2.5	1.7	1.25
HF572913, *Pseudomonas* sp. SMX321	*Proteobacteria*	2.5	2.5	1.7
HE985241, *Pseudomonas* sp. SMX330	*Proteobacteria*	2.5	1.7	1.25
HF571533, *Pseudomonas* sp. SMX331	*Proteobacteria*	2.5	1.7	1.25
HF571535, *Pseudomonas* sp. SMX344	*Proteobacteria*	2.5	1.7	1.25
HF571536, *Pseudomonas* sp. SMX345	*Proteobacteria*	2.5	1.25	1.25
HF571534, *Variovorax* sp. SMX332	*Proteobacteria*	2.5	1.7	1.25

Therefore, sorption onto biomass or other materials was shown to be negligible and SMX removal must be due to biodegradation.

Photodegradation was excluded by performing all experiments in the dark.

To characterize biodegradation ability and rate and evaluate an optimal nutrient environment for SMX utilization of the isolated and identified 9 pure cultures, subsequent experiments were performed. In the presence of readily degradable nutrients (Figure 7.3, Figure 7.4) SMX was faster biodegraded compared to setups with SMX as sole carbon/nitrogen source (Figure 7.5). 54 setups (three media for each of the 9 cultures in duplicate setups) with different nutrient compositions were set up and SMX biodegradation rates were evaluated using UV-AM values (Table 7.2). Different SMX biodegradation patterns were observed indicating that the presence or absence of readily degradable and complex nutrients significantly influenced biodegradation. R2A-UV media were sampled once a day as it was assumed that biodegradation might be faster compared to the other two nutrient-poor media. Biodegradation rates of 2.5 mg L^{-1} d^{-1} were found for all nine species not showing any different biodegradation behaviors or patterns (Figure 7.3 A). Biomass growth affected background absorbance that was increased with cell density but UV-AM could still be applied to monitor biodegradation as the background was still low enough.

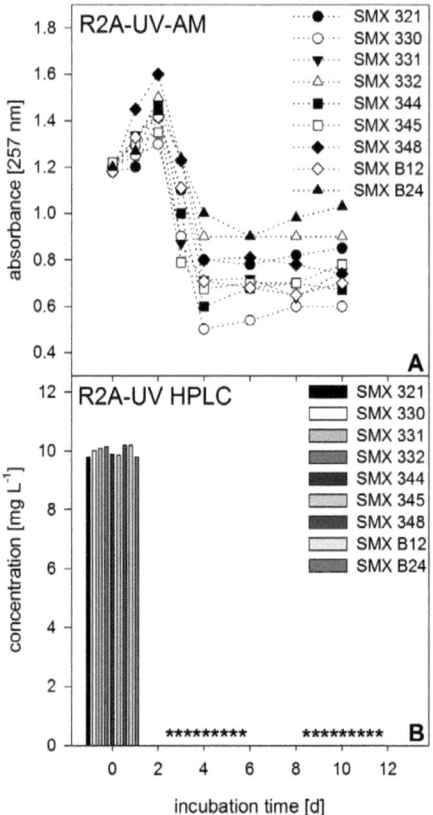

Figure 7.3 A and B – A) Aerobic SMX biodegradation patterns of pure cultures in R2A-UV media
measured with UV-AM. Initial SMX concentration of 10 mg L^{-1}. B) LC-UV analyses of SMX concentrations within the nine pure cultures in R2A-UV media performed at experimental startup, after 4 and 10 days to verify the results of UV-AM. Asterisks indicate measured values below limit of detection. Shown are mean SMX absorbance values of duplicate experiments. Standard deviations were too low to be shown (<1%).

Chapter 7 – Sulfamethoxazole biodegradation by pure cultures

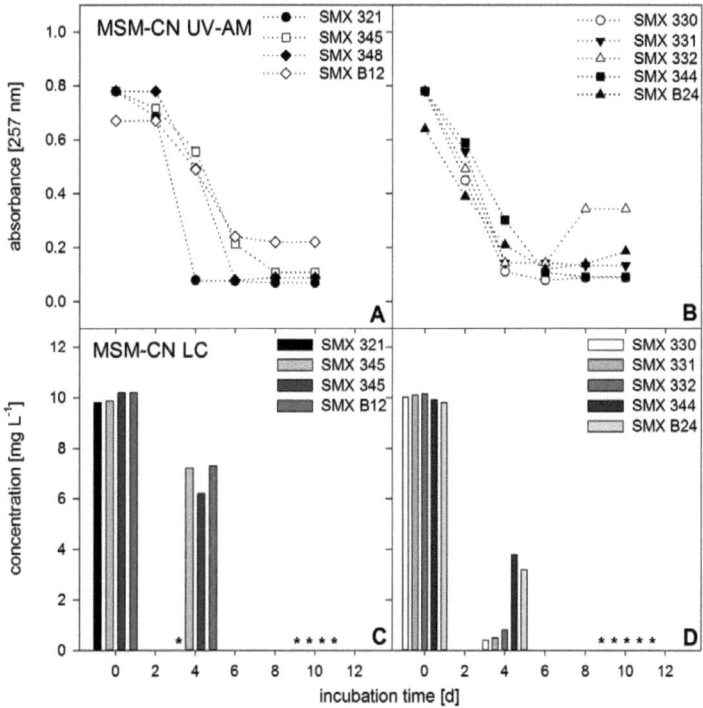

Figure 7.4 A, B, C and D – A, B) Aerobic SMX biodegradation patterns of pure cultures in MSM-CN
media measured with UV-AM. Initial SMX concentration of 10 mg L^{-1}. C, D) LC-UV analyses of SMX concentrations in the used pure cultures in MSM-CN. Determination was performed at experimental startup, after 4 and 10 days to verify UV-AM values. Asterisks indicate measured values below limit of detection. Shown are mean values of SMX absorbance in duplicate experiments. Standard deviations were too low to be shown (<1%).

In MSM-CN (Figure 7.4), offering only specific C- and N-sources (acetate and ammonium nitrate), the biodegradation rates ranged from 1.25 to 2.5 mg L^{-1} d^{-1} (standard deviations between the duplicate setups were below 1%) showing clear differences for the different species, even for the five *Pseudomonas* spp.. While

Pseudomonas sp. SMX321 biodegraded SMX with 2.5 mg L^{-1} d^{-1}, *Pseudomonas* sp. SMX344 acchieved only 1.25 mg L^{-1} d^{-1}. The same effect was found for the two *Microbacterium* spp.. SMXB12 showed 1.7 mg L^{-1} d^{-1} while SMX348 only showed 1.25 mg L^{-1} d^{-1}.

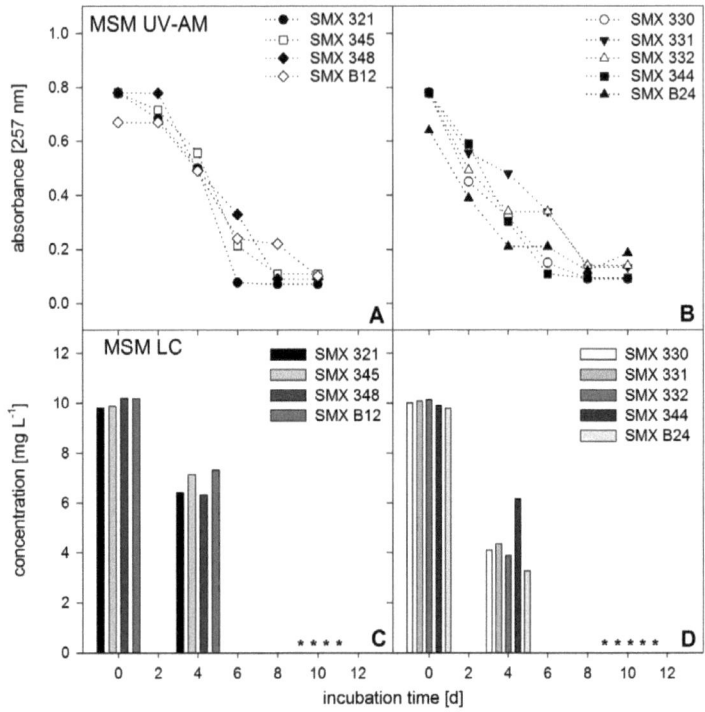

Figure 7.5 A, B, C and D – A, B) Aerobic SMX biodegradation patterns of pure cultures in MSM media
measured with UV-AM. Initial SMX concentration of 10 mg L^{-1}. C, D) LC-UV analyses of SMX concentrations in the pure cultures in MSM at experimental startup, after 4 and 10 days to validate UV-AM. Asterisks indicate measured values below limit of detection. Shown are mean values of SMX absorbance in duplicate experiments. Standard deviations were too low to be shown (<1%).

Biodegradation pattern in MSM-CN of four isolates (SMX321, 345, 348 and B12) revealed a short lag phase of two days with no SMX removal (Figure 7.4 A) while the other five were able to biodegrade SMX already after two days and showed a constant SMX removal during cultivation (Figure 7.4 B). In MSM (Figure 7.5), with SMX as sole C- and N-source, the removal rate of SMX was even lower. Biodegradation rates of 1.0 mg L^{-1} d^{-1} were found for *Brevundimonas* sp. SMXB12 while *Pseudomonas* sp. SMX321 showed 1.7 mg L^{-1} d^{-1}. All other species showed removal rates of 1.25 mg L^{-1} d^{-1}. These experiments with SMX as sole C/N-source demonstrated that it could serve as nutrient source but with up to 2.5-fold reduced biodegradation rates. Biodegradation pattern in MSM was similar to that in MSM-CN with a lag phase of two days for the four isolates SMX321, 345, 348 and B12 (Figure 7.5 A) and no lag phase for the isolates SMX 330, 331, 332, 344, and B24 starting to utilize SMX already after two days (Figure 7.5 B).

In general, it was found that the five *Pseudomonas* spp. and the two *Microbacterium* spp. did not show the same biodegradation behavior. At least one member of each group always showed a lag phase while the other immediately started SMX biodegradation.

As UV-AM revealed sufficient to monitor SMX biodegradation (Figure 7.1 and Table 7.1) LC-UV measurements were performed at the start of the experiment, after four and 10 days as control measurements (Figure 7.3 B, Figure 7.4 C and D, Figure 7.5 C and D).

LC-UV showed that in R2A-UV all cultures removed 10 mg L^{-1} SMX in 4 days (Figure 7.3 B) while in MSM-CN only *Pseudomonas* sp. SMX321 removed all SMX within 4 days (Figure 7.4 C). The remaining 8 cultures still showed residual SMX concentrations from

0.4 to 7.3 mg L^{-1} and complete SMX elimination was achieved only at day 10 (Figure 7.4 C and D).

In MSM after 4 days, SMX was still present in all nine cultures in concentrations above 3.6 mg L^{-1} and only after 10 days SMX was below the limit of detection (Figure 7.5 C and D). LC-UV values could be compared to UV-AM values and approved this simple approach to be applicable for screening SMX biodegradation.

7.4 DISCUSSION

This study focused on the cultivation of pure culture SMX biodegrading organisms to perform specific biodegradation experiments. It is known that cultivation, especially on solid media, is affected with the problem described as "viable but non cultivable" (VBNC) [201, 202]. Solid media being implicitly required for the isolation of pure cultures is for sure limited in its cultivation efficiency mainly due to reduced water content and different or inappropriate nutrient conditions. Only a low percentage of around 1% of the active organisms in environmental samples [203] and around 15% from activated sludge can be cultivated [204, 205] but in this study, 9 different isolates out of 110 pure cultures were obtained showing SMX biodegradation. This quite high percentage of almost 10% was only possible with a two-step SMX-acclimation experiment that was conducted to increase the chance to cultivate SMX biodegrading organisms by applying a strong selective pressure using 10 mg L^{-1} SMX in the media. Furthermore, R2A medium that is known to work well for isolation of aquatic organisms [139], was applied for the cultivation of bacteria being assumed to be at least SMX-resistant when growth was observed on SMXreinforced R2A. However, a lot more organisms compared to those cultivated in this study, might be present in activated sludge capable of SMX biodegradation. These VBNCs might be taxonomically characterized by culture-independent methods [206, 207]. However, for our focus on linking biodegradation patterns, rates and nutrient utilization to specific species these methods were not feasible. Only with actively biodegrading pure cultures a clear and precise coherence between SMX biodegradation and taxonomically identified species is possible. As a final goal, pure cultures would allow to analyze species-specific biodegradation products and thus determine potential SMX biodegradation pathways. Applying that knowledge to WWTP techniques would provide a strategy to selectively enhance biodegrading species in activated sludge systems improving and stabilizing SMX removal efficiency.

Therefore, phylogenetic identification of potential SMX biodegrading species is implicitly required. As shown in this study, five of the nine SMX biodegrading species found belonged to the genus *Pseudomonas* confirming this group to play an important role for the biodegradation of micropollutants. This was reported for e.g. acetaminophen or chlorinated compounds by many other studies [52, 158, 208]. Additionally, two isolates SMXB24 and SMX348 were identified as *Microbacterium* sp.. It was shown that *Microbacterium* sp. SMXB24 is closely related to *Microbacterium* sp. 7 1K, an organism that was found to be related with phytoremediation. The second *Microbacterium* sp. SMX348 is closely related to *Microbacterium* sp. BR1 which was isolated from an acclimated SMX biodegrading membrane bioreactor, supporting this species' crucial role for the biodegradation of SMX [104]. In addition, the general potential of different *Microbacteria* species for the biodegradation of xenobiotic compounds has been highlighted in the literature [209, 210]. Also *Variovorax paradoxus*, closely related to the isolated *Variovorax* sp. SMX332, is reported to be capable of biodegrading a large variety of pollutants including sulfolene and other heterocyclic compounds [211]. Therefore, it seems likely that the isolated *Variovorax* sp. SMX332 might also be able to biodegrade SMX. Finally, some literature data also exist for the group *Brevundimonas* spp. indicating that these organisms might play a role in the removal of antibiotics [212].

Taxonomic identification was followed by observing influences on biodegradation rate and efficiency due to the availability of nutrients. Biodegradation rates decreased with reduced nutrient content from the complex R2A-UV over nutrient-poor MSM-CN and MSM media and more time was needed to remove SMX. MSM media contained SMX as sole carbon and nitrogen source at a concentration of 10 mg L^{-1} and thus provided just around 4.8 mg L^{-1} carbon and 1.7 mg L^{-1} nitrogen. These conditions, with SMX being the only nutrient in MSM, showed an effect on biodegradation and reduced removal efficiency but demonstrated the organisms' ability to utilize SMX as sole nutrient and/or energy source. However, this indicates that complex nutrients and higher nutrient concentrations seem to have a positive effect on biodegradation due to co-metabolic [105] or diauxic effects [213] as the very high SMX removal rates of 2.5 mg L^{-1} d^{-1} confirmed. They were significantly higher than 0.0079 mg L^{-1} d^{-1} found in a previous study [51]. In general, SMX biodegradation might be based more on a diauxic process, i.e. readily degradable nutrients are used up first followed by SMX utilization, rather than real co-metabolism, i.e. two substrates are used up in parallel when provided together, as experiments with R2A-UV media showed. A strong increase in UV-AM, attributed to biomass growth due to a fast nutrient consumption provided by the complex R2A-UV media, was followed by a rapid SMX elimination. This increase was not observed in MSM-CN or MSM as the nutrients concentrations were too low to foster excessive biomass growth. Even at low cell densities SMX was rapidly removed indicating that biomass concentration is not as important as cellular activity. Therefore, the higher removal rates in presence of sufficient nutrients also showed that SMX biodegradation was a rapid and complex metabolic process.

Therefore, information about the biodegradation potential of the isolated bacterial strains with respect to the availability of nutrients might increase the elimination efficiency in WWTPs as the treatment process could be specifically adapted to the needs of the biodegrading species.

For future research, the availability of isolated species will allow screening for biodegradation intermediates and/or stable metabolites and determination of species-specific biodegradation pathways. To date only few data on SMX metabolites such as 3-amino-5-methyl-isoxazole found in SMX degrading ASCs [50] and hydroxy-N-(5-methyl-1,2-oxazol-3-yl)benzene-1-sulfonamide detected in an SMX degrading consortium of fungi and

Rhodococcus rhodochrous exists [105]. Further research is also needed to screen for the nutrient influence on metabolite formation, i.e. if the isolated pure cultures produce different metabolites due to changing nutrient conditions.

7.5 CONCLUSIONS

The present work focused on nine different bacteria species capable of Sulfamethoxazole (SMX) biodegradation isolated from SMX-acclimated ASCs. Initially 110 pure cultures were screened for their SMX biodegradation potential and nine observed biodegradation potential. These were identified via 16S rRNA sequencing revealing five *Pseudomonas* spp., one *Brevundimonas* sp., one *Variovorax* sp. and two *Microbacterium* spp.. Thus, seven species belonged to the phylum *Proteobacteria* and two to *Actinobacteria*. Incubation in media containing 10 mg L^{-1} SMX and different concentrations of carbon and nitrogen showed different biodegradation patterns with respect to media composition and bacterial species. Biodegradation occurred very fast with 2.5 mg L^{-1} d^{-1} SMX being biodegraded in all pure cultures in complex nutrient R2A-UV media under aerobic conditions and room temperature. Biodegradation rates were lower for setups with SMX provided as co-substrate together with carbon/nitrogen at a ratio of DOC: N – 33: 1 with rates ranging from 1.25 to 2.5 mg L^{-1} d^{-1}. Media containing SMX as sole carbon and nitrogen source showed the organisms' ability to use SMX as the only nutrient source but biodegradation rates decreased to 1.0 – 1.7 mg L^{-1} d^{-1}. The different species showed specific biodegradation rates and behaviours at various nutrient conditions. Readily degradable energy sources seemed crucial for efficient SMX biodegradation for all species.

Chapter 8

Conclusions and Outlook

8.1 CONCLUSIONS

8.1.1 Efficient Screening for Biodegradation in Laboratory Setups

Generally, biodegradation experiments require a high number of different setups to evaluate various parameters and study biodegradation under different conditions, e.g. nutrients, biomass concentration. Therefore, an easy-to-perform and cost-efficient methodology is inevitable and thus a UV-absorbance-based measurement (UV-AM) system was established. This methodology enabled further experiments revealing biodegradation patterns that contribute to the more complete understanding of the processes during SMX and BTs biological removal. This system allowed analyzing both benzotriazole (1-H-benzotriazole, 4-tolyltriazole and 5-tolyltriazole) and sulfamethoxazole (SMX, an antibiotic) within a large number of biodegradation experiments. These experiments would normally require time-intensive and money-consuming LC/GC measurements but are, in case of many laboratory setups, not always needed. In lab scale setups, often high concentrations of the xenobiotica are used to apply a high selective pressure on the microbial communities/pure cultures to achieve a fast acclimation. These concentrations could be monitored by a simple UV-absorbance measurement (UV-AM) that was developed and validated for screening a large number of setups. This technique required hardly any preparation and significantly less time and money compared to LC/GC methods. UV-AM was evaluated by comparing its measured values with LC-UV and GC-MS/MS results. Furthermore, its application for monitoring and screening unknown activated sludge communities (ASCs) and mixed pure cultures has been approved to detect xenobiotica biodegradation of benzotriazole (BTri), 4- and 5-tolyltriazole (4-TTri, 5-TTri) as well as SMX. In laboratory setups, xenobiotic concentrations above 1.0 mg L^{-1} without any enrichment or preparation were detectable after optimization of the method. UV-AM did not require any specific preparatory work and could be performed in 96 or even 384 well plate formats thus significantly increasing the number of possible parallel setups and screening efficiency whereas analytic and laboratory costs were reduced to a minimum. With this technique, further experiments were possible to reveal specific biodegradation patterns of the used ASCs/pure cultures with respect to the applied xenobiotica.

8.1.2   Benzotriazoles – Monitoring

Benzotriazole (BTri), 4-methyl-(4-TTri) and 5-methyl-benzotriazole (5-TTri), summarized as BTs, were monitored over one year in three wastewater treatment plants (WWTPs MBR-MH, CAS-E and CAS-M) operated with different treatment regimes, i.e. a membrane bioreactor (MBR-MH), a conventional activated sludge plant with intermittent nitrification/ denitrification (CAS-E) and a two stage activated sludge plant (CAS-M). These WWTPs were monitored for their BTs influent/effluent concentrations, removal efficiencies of the treatment stages, and their impact on the receiving rivers. Furthermore, as the three WWTPs receive different wastewater volumes from industry, inferences can be drawn regarding the influence of industry on BTs concentrations.

With a mean of 75%, 5-TTri was removed best in all three WWTPs and that removal occurred mainly in the aeration tanks, probably due to biodegradation. In contrast, BTri, with a mean total removal of 45% and 4-TTri, with a mean total removal of only 15% showed significantly lower removal efficiencies. Their removal took place across all the different wastewater treatment stages such as primary clarifier, aeration tank and sedimentation tank. Both compounds were equally removed biologically, i.e. biodegradation, and abiotically, probably by photodegradation. High removal fluctuations over the four seasons of the year were observed in all three WWTPs for the BTs but lowest removal occurred during winter.

Additionally, measurements of the receiving rivers up- and downstream the WWTPs showed that they constituted a point source for BTs in the aquatic environment. Significantly increased BTs concentrations were detected for MBR-MH (total mean 3.58 µg L^{-1}) and CAS-M (total mean 1.1 µg L^{-1}) downstream the WWTP while the rivers upstream the WWTPs showed hardly any BTs contamination (total mean <0.4 µg L^{-1}). In contrast, WWTP CAS-E effluents only slightly increased the river's downstream concentrations up to 0.64 µg L^{-1} as the receiving river was already contaminated with BTs (total mean 0.50 µg L^{-1}) due to diffuse entries from hydropower. Regarding the downstream concentrations of the individual BTs substances, BTri was found ranging from 0.43 to 1.66 µg L^{-1}, 5-TTri from 0.03 to 0.16 µg L^{-1} and 4-TTri from 0.18 to 1.76 µg L^{-1}.

In general, 5-TTri showed a good removal during wastewater treatment and was thus detected at lower effluent concentrations compared to BTri and 4-TTri that showed a minor removal in WWTPs resulting in higher effluent concentrations. Therefore, these two compounds have the potential to accumulate in the receiving rivers, as their self-purification potential presumably is not enough to eliminate these compounds. However, also the better removed 5-TTri was not completely eliminated and occurred in the effluents implying the need for further research on specifically optimizing BTs removal efficiency during wastewater treatment.

8.1.3 Benzotriazoles - Biodegradation

The monitoring results indicated that biodegradation might be the preferential removal mechanism for 5-TTri and also contribute to BTri and 4-TTri removal. Therefore, laboratory experiments were conducted to evaluate the impact of biodegradation on BTs removal. For that purpose, three ASCs, taken from the monitored three WWTPs were evaluated for their biodegradation potential on benzotriazole (BTri), 4-methyl-benzotriazole (4-TTri) and 5-methyl-benzotriazole (5-TTri). In addition, their biodegradation ability under aerobic, denitrifying, sulfate reducing and anaerobic conditions as well as under supply of different nutrients was evaluated.

All three ASCs were able to biodegrade up to 30 mg L^{-1} 5-TTri and BTri under aerobic conditions within up to 7 and 49 days, respectively, but showed no removal under denitrifying, sulfate reducing or anaerobic conditions. In contrast to these results, 4-TTri was refractory under all conditions tested explaining why 4-

TTri also showed the lowest elimination during wastewater treatment.

Furthermore, significant differences were observed for BTri biodegradation with non-acclimated sludge: ASC from MBR-MH, CAS-E and CAS-M needed 21d, 41d and 49 d, respectively, to remove BTri. After acclimation, all three ASCs removed BTri within 7 days. In addition, different carbon and nitrogen concentrations revealed that nitrogen was implicitly required for biodegradation while carbon showed no such effect. In presence of additionally supplied nitrogen, biodegradation was faster and more efficient than with only carbon sources being present. As a strong nitrogen dependency during BTri biodegradation was observed it might be that 5-TTri biodegradation, although fast compared to BTri, could also be improved by nitrogen supply. As 5-TTri, although observing a better biodegradation compared to BTri, is not completely eliminated during wastewater treatment, further improvements could achieve biodegradation rates being high enough to ensure an almost complete removal. Therefore, additional experiments regarding 5-TTri biodegradation enhancement were set up to evaluate the optimal conditions for 5-TTri removal.

As 5-tolyltriazole (5-TTri) might, despite its good biodegradability, have a detrimental impact on aquatic systems an acute need to reduce the concentration of 5-TTri in the effluents of WWTPs is given. Therefore, further research focusing on the enhancement of 5-TTri biodegradation through acclimation of the ASCs to high concentrations of 5-TTri as well as by supplying additional nutrients was required. The results might also help to improve the biological removal of BTri. For these experiments, the ASCs were pre-screened in presence of 20 mg L^{-1} 5-TTri. Biodegrading biomass was subsequently used to inoculate a following generation that finally led to nine ASC generations. ASC generation two was characterized by a lag phase of five days without biodegradation that was followed by rapid removal of 5-TTri. ASC generation three behaved similarly but exhibited a lag phase of four days. The following generations four to nine were, in contrast to generations two and three, able to utilize 5-TTri immediately after inoculation and showed high biodegradation rates ranging from 3.3 to 5.2 mg L^{-1} d^{-1}. An additional experiment performed with the supernatant from centrifuged activated sludge that was used to simulate the complex nutrient conditions in wastewater, demonstrated that this sludge supernatant (SS) significantly enhanced biodegradation. The natural nutrients contained in SS resulted in removal rates ranging from 3.2 to 5.0 mg L^{-1} d^{-1} without prior ASC acclimation. The control group without SS showed significantly lower biodegradation rates of ≤ 2.2 mg L^{-1} d^{-1}. These SS experiments indicated that 5-TTri removal might be strongly dependent on nitrogen supply, a behavior already observed for BTri. Presumably biodegradation begins by benzene ring cleavage and is necessitating nitrogen-supply. Subsequent experiments performed with three nitrogen species (NH_4NO_3, NH_4Cl and $NaNO_3$) also significantly enhanced biodegradation. These results further strengthened the theory of nitrogen being crucial for 5-TTri biodegradation. All supplied nitrogen species enhanced biodegradation and enabled the ASC to utilize 5-TTri without the need for acclimation as was already observed for SS-supplied setups.

Results from biodegradation experiments with ASCs indicated that 5-TTri and probably BTri might be removed in a concerted syntrophic metabolic reaction involving different bacteria, as it was not possible to obtain pure BTs biodegrading bacterial cultures, even after intensive effort. To obtain an idea, which organisms might be involved in biological BTs removal processes, an intensive characterization approach focusing on 5-TTri biodegrading ASC including denaturing gradient gel electrophoresis (DGGE), metagenomic and metatranscriptomic analyzes, was adopted. DGGE analysis revealed a low biodiversity in the biodegradation setups consisting of mainly four dominant species, i.e. *Aminobacter* spp., *Flavobacterium* spp., *Hydrogenophaga* spp. and *Pseudomonas* spp.. In contrast, metagenomic analysis showed a high diversity across the different biodegradation setups but only few organisms with a high abundance. The most prominent species in the non-biodegrading setup were *Mesorhizobium* spp. and *Hydrogenophaga* spp. while in the biodegrading setups *Acidovorax* spp., *Hydrogenophaga* spp. and *Pseudomonas* spp. were found in highest abundance. *Hydrogenophaga* spp. showed a decreased activity in the non-biodegrading setup but a significantly increased activity in the biodegrading one. *Pseudomonas* spp. showed a very high overall activity. However, it might be that *Pseudomonas* spp. do not biodegrade 5-TTri alone and the formation of a biodegrading ASC is required. The third dominant genus, i.e. *Acidovorax*, also showed a high activity in the biodegrading setups indicating that these genus might also contribute to 5-TTri biodegradation. Their high metabolic activity might allow the organisms to easily establish a functioning biodegrading ASC.

To finally conclude, this research project evaluated the biodegradation of three benzotriazoles together with the antibiotic sulfamethoxazole with respect to different redox and nutrient conditions. Therefore, experiments with ASCs and pure cultures were conducted and demonstrated that BTri and 5-TTri as well as sulfamethoxazole were biodegradable under aerobic conditions but not under denitrifying, sulfate reducing or anaerobic conditions. 4-methyl-benzotriazole was biologically stable under all applied conditions. An additional monitoring study over one year in three different wastewater treatment plants showed a similar benzotriazoles' behavior than observed in the laboratory experiments but a strong fluctuation regarding removal efficiencies.

8.1.4 Sulfamethoxazole – Identification of Pure, Biodegrading Bacterial Cultures

In this experimental approach, SMX (sulfonamide antibiotic) biodegradation was evaluated. SMX biodegradation by ASCs is still only partly understood. Thus, extensive investigations were performed and finally resulted in nine different bacteria species, capable of SMX biodegradation, isolated from ASCs. These isolates allow to specifically correlate biodegradation pattern with the bacterial species and performing specific optimizations. Therefore, all nine isolates were subsequently identified via 16S rRNA gene sequencing and revealed five *Pseudomonas* spp., one *Brevundimonas* sp., one *Variovorax* sp. and two *Microbacterium* spp.. Thus, seven species belonged to the phylum *Proteobacteria* and two to *Actinobacteria*. These cultures were incubated in media containing 10 mg L^{-1} SMX and different nutrients, i.e. complex nutrients, specific nutrients carbon (sodium acetate) and nitrogen (ammonium nitrate), but also mineral salt medium without nutrients. Different biodegradation patterns were revealed with respect to media composition and bacteria species. It occurred very fast with 2.5 mg L^{-1} d^{-1} SMX being biodegraded in all pure cultures in complex nutrient R2A-UV media under aerobic conditions and room temperature. However, different and reduced biodegradation rates were observed for setups with SMX provided as co-substrate together with specific carbon (sodium acetate) and nitrogen (ammonium nitrate) supply at a ratio of DOC: N – 33: 1 with rates ranging from 1.25 to 2.5 mg L^{-1} d^{-1}. Media containing only SMX as carbon and nitrogen source showed the organisms' ability to use SMX as sole nutrient and energy source where biodegradation rates decreased to 1.0 – 1.7 mg L^{-1} d^{-1}. Species-specific biodegradation rates and behaviors at various nutrient conditions occurred and confirmed that throughout readily degradable energy sources were crucial for efficient SMX biodegradation. Thus, pure cultures were shown to be capable of SMX biodegradation and experiments with different nutrients showed their impact on biodegradation rates. Such experiments are hardly possible with ASCs as their composition is unknown and nutrients can emerge from the ASC itself. Isolates are thus inevitable for specific biodegradation experiments.

8.2 OUTLOOK

This work covered the large topic of biodegradation with the focus on the benzotriazoles 1-H-benzotriazole (BTri), 4-methyl- and 5-methyl-benzotriazole (4-TTri, 5-TTri), as well as sulfamethoxazole (SMX). It was shown that ASCs, under the premise of the right conditions, were able to remove all mentioned compounds except 4-methyl-benzotriazole. Therefore, further research is implicitly required on the removal of 4-TTri by means of abiotic processes, i.e. advanced oxidation processes, ozonation or UV-treatment, as this compound poses a high risk to accumulate in the aquatic environment. BTri showed a weak and inconsistent removal during wastewater treatment and might be prone to accumulation while 5-TTri and SMX observed a good biodegradation potential. Therefore, for these two compounds, additional treatment steps, e.g. a fourth treatment stage, might not be required. However, also 5-TTri and SMX are not completely removed and adaptations and improvements in the existing wastewater treatment systems are necessary to ensure a consistent and complete biological removal. Nevertheless, the optimal way to reduce these compounds would be to minimize their application and/or, especially for BTri and 4-TTri, their replacement. 5-TTri, as potent in corrosion protection as the other two compounds, could easily replace them.

Additional efforts are surely required to evaluate further aspects regarding biodegradation. In case of the benzotriazoles, experiments to determine optimal biodegradation conditions for ASCs are still necessary and might finally result in an appropriate wastewater treatment strategy. Furthermore it is still unknown which effects may lead to the biological discrimination of 4-TTri that exhibits almost the same chemical characteristics as 5-TTri. Analyzes regarding the enzymes involved in 5-TTri biodegradation metabolism might help to solve that problem. Additionally, such research would also lead to the determination of enzymatic degradation pathways and to the identification of transformation products occurring during BTs biodegradation. Therefore, pure bacterial cultures are inevitable and further efforts have to be undertaken to finally gain biodegrading isolates. A first step, the characterization of 5-TTri biodegrading ASC with next-generation-sequencing was already accomplished during this project but further experiments must aim on finally resolving the biodegrading community composition.

The same aspects should be considered for SMX biodegradation. As already pure cultures are available, it might be possible to determine metabolic enzymatic pathways explaining the underlying cellular processes. Knowledge of these pathways would enable specific metabolic engineering approaches to optimize biodegradation or use direct enzyme-driven setups for a cell-independent and cell-free removal strategy for SMX during wastewater treatment whenever necessary.

In general, for both BTs and SMX, further research is required on transformation products arising from biodegradation. Possible transformation products need to be characterized regarding their aquatic toxicity and persistency to finally assess the full impact on aquatic systems of BTs and SMX biodegradation. The formation of stable transformation products might be possible and the application of advanced oxidation processes as the fourth treatment stage could be a viable option to finally eliminating BTs and SMX from WWTPs effluents.

9 REFERENCES

1. **Pschyrembel W.** 2013. Pschyrembel Klinisches Wörterbuch, p. 2320, Pschyrembel, 2013 ed, vol. 264. De Gruyter.
2. **Schwarzenbach RP, Escher BI, Fenner K, Hofstetter TB, Johnson CA, von Gunten U, Wehrli B.** 2006. The Challenge of Micropollutants in Aquatic Systems. Science **313**:1072-1077.
3. **Fatta-Kassinos D, Bester K.** 2010. Xenobiotics in the urban water cycle: mass flows, environmental processes, mitigation and treatment strategies, vol. 16. Springer.
4. **Ding C, He J.** 2010. Effect of antibiotics in the environment on microbial populations. Appl. Microbiol. Biotechnol. **87**:925-941.
5. **Dann AB, Hontela A.** 2011. Triclosan: environmental exposure, toxicity and mechanisms of action. J. Appl. Toxicol. **31**:285-311.
6. **Keiter S.** 2013. Long-term effects and chemosensitizing potential of perfluorinated chemicals (PFCs) in zebrafish (Danio rerio).
7. **Köck-Schulmeyer M, Villagrasa M, López de Alda M, Céspedes-Sánchez R, Ventura F, Barceló D.** 2013. Occurrence and behavior of pesticides in wastewater treatment plants and their environmental impact. Sci. Total Environ. **458–460**:466-476.
8. **Kulkarni D, Gergs A, Hommen U, Ratte H, Preuss T.** 2013. A plea for the use of copepods in freshwater ecotoxicology. Environ Sci Pollut R **20**:75-85.
9. **Preiss A, Berger-Preiss E, Elend M, Gerling S, Kühn S, Schuchardt S.** 2012. A new analytical approach for the comprehensive characterization of polar xenobiotic organic compounds downgradient of old municipal solid waste (MSW) landfills. Anal. Bioanal. Chem. **403**:2553-2561.
10. **Miège C, Choubert JM, Ribeiro L, Eusèbe M, Coquery M.** 2009. Fate of pharmaceuticals and personal care products in wastewater treatment plants – Conception of a database and first results. Environ. Pollut. **157**:1721-1726.
11. **Ort C, Lawrence MG, Rieckermann J, Joss A.** 2010. Sampling for Pharmaceuticals and Personal Care Products (PPCPs) and Illicit Drugs in Wastewater Systems: Are Your Conclusions Valid? A Critical Review. Environ. Sci. Technol. **44**:6024-6035.
12. **Omole D, Adewumi I, Longe E, Ogbiye A.** 2012. Study of Auto Purification Capacity of River Atuwara in Nigeria. International Journal of Engineering and Technology **2**:229-235.
13. **Luo L, Wang XC, Guo W, Ngo HH, Chen Z.** 2012. Impact assessment of excess discharges of organics and nutrients into aquatic systems by thermodynamic entropy calculation. J. Environ. Manage. **112**:45-52.
14. **Jelić A, Petrović M, Barceló D.** 2012. Pharmaceuticals in Drinking Water, p. 47-70. *In* Barceló D (ed.), Emerging Organic Contaminants and Human Health. Springer Berlin Heidelberg.
15. **Loraine GA, Pettigrove ME.** 2005. Seasonal Variations in Concentrations of Pharmaceuticals and Personal Care Products in Drinking Water and Reclaimed Wastewater in Southern California. Environ. Sci. Technol. **40**:687-695.
16. **Hoa PTP, Managaki S, Nakada N, Takada H, Shimizu A, Anh DH, Viet PH, Suzuki S.** 2011. Antibiotic contamination and occurrence of antibiotic-resistant bacteria in aquatic environments of northern Vietnam. Sci. Total Environ. **409**:2894-2901.
17. **Loos R, Locoro G, Comero S, Contini S, Schwesig D, Werres F, Balsaa P, Gans O, Weiss S, Blaha L, Bolchi M, Gawlik BM.** 2010. Pan-European survey on the occurrence of selected polar organic persistent pollutants in ground water. Water Res. **44**:4115-4126.

18. **Davies J, Spiegelman GB, Yim G.** 2006. The world of subinhibitory antibiotic concentrations. Curr. Opin. Microbiol. **9:**445-453.
19. **Kümmerer K.** 2009. Antibiotics in the aquatic environment--a review--part I. Chemosphere **75:**417-434.
20. **Allen HK, Donato J, Wang HH, Cloud-Hansen KA, Davies J, Handelsman J.** 2010. Call of the wild: antibiotic resistance genes in natural environments. Nature Reviews Microbiology **8:**251-259.
21. **Biyela P, Lin J, Bezuidenhout C.** 2004. The role of aquatic ecosystems as reservoirs of antibiotic resistant bacteria and antibiotic resistance genes. Water Sci. Technol. **50:**45-50.
22. **Fan C, Lee PH, Ng W, Alvarez-Cohen L, Brodie E, Andersen G, He J.** 2009. Influence of trace erythromycin and erythromycin-H2O on carbon and nutrients removal and on resistance selection in sequencing batch reactors (SBRs). Appl. Microbiol. Biotechnol. **85:**185-195.
23. **Sarmah AK, Meyer MT, Boxall ABA.** 2006. A global perspective on the use, sales, exposure pathways, occurrence, fate and effects of veterinary antibiotics (VAs) in the environment. Chemosphere **65:**725-759.
24. **Kidd KA, Blanchfield PJ, Mills KH, Palace VP, Evans RE, Lazorchak JM, Flick RW.** 2007. Collapse of a fish population after exposure to a synthetic estrogen. Proceedings of the National Academy of Sciences **104:**8897-8901.
25. **Schaar H, Clara M, Gans O, Kreuzinger N.** 2010. Micropollutant removal during biological wastewater treatment and a subsequent ozonation step. Environ. Pollut. **158:**1399-1404.
26. **Stasinakis AS, Thomaidis NS, Arvaniti OS, Asimakopoulos AG, Samaras VG, Ajibola A, Mamais D, Lekkas TD.** 2013. Contribution of primary and secondary treatment on the removal of benzothiazoles, benzotriazoles, endocrine disruptors, pharmaceuticals and perfluorinated compounds in a sewage treatment plant. Sci. Total Environ. **463–464:**1067-1075.
27. **Munir M, Wong K, Xagoraraki I.** 2011. Release of antibiotic resistant bacteria and genes in the effluent and biosolids of five wastewater utilities in Michigan. Water Res. **45:**681-693.
28. **Noppe H, Verheyden K, Gillis W, Courtheyn D, Vanthemsche P, De Brabander HF.** 2007. Multi-analyte approach for the determination of ng L−1 levels of steroid hormones in unidentified aqueous samples. Anal. Chim. Acta **586:**22-29.
29. **Sumpter JP, Jobling S.** 1995. Vitellogenesis as a biomarker for estrogenic contamination of the aquatic environment. Environ. Health Perspect. **103:**173.
30. **Lee W, Kang C-W, Su C-K, Okubo K, Nagahama Y.** 2012. Screening estrogenic activity of environmental contaminants and water samples using a transgenic medaka embryo bioassay. Chemosphere **88:**945-952.
31. **Corcoran J, Winter MJ, Tyler CR.** 2010. Pharmaceuticals in the aquatic environment: a critical review of the evidence for health effects in fish. Crit. Rev. Toxicol. **40:**287-304.
32. **Joss A, Keller E, Alder AC, Göbel A, McArdell CS, Ternes T, Siegrist H.** 2005. Removal of pharmaceuticals and fragrances in biological wastewater treatment. Water Res. **39:**3139-3152.
33. **Omil F, Suárez S, Carballa M, Reif R, Lema J.** 2010. Criteria for Designing Sewage Treatment Plants for Enhanced Removal of Organic Micropollutants, p. 283-306. *In* Fatta-Kassinos D, Bester K, Kümmerer K (ed.), Xenobiotics in the Urban Water Cycle, vol. 16. Springer Netherlands.
34. **Weiss S, Reemtsma T.** 2008. Membrane bioreactors for municipal wastewater treatment - a viable option to reduce the amount of polar pollutants discharged into surface waters? Water Res. **42:**3837-3847.

35. **Cirja M, Ivashechkin P, Schäffer A, Corvini PX.** 2008. Factors affecting the removal of organic micropollutants from wastewater in conventional treatment plants (CTP) and membrane bioreactors (MBR). Reviews in Environmental Science and Bio/Technology **7**:61-78.
36. **Sipma J, Osuna B, Collado N, Monclús H, Ferrero G, Comas J, Rodriguez-Roda I.** 2010. Comparison of removal of pharmaceuticals in MBR and activated sludge systems. Desalination **250**:653-659.
37. **Pollice A, Laera G, Saturno D, Giordano C.** 2008. Effects of sludge retention time on the performance of a membrane bioreactor treating municipal sewage. J. Membr. Sci. **317**:65-70.
38. **Fane AG.** 2002. Membrane bioreactors: design and operational options. Filtration & Separation **39**:26-29.
39. **Cicek N, Macomber J, Davel J, Suidan M, Audic J, Genestet P.** 2001. Effect of solids retention time on the performance and biological characteristics of a membrane bioreactor. Water Sci. Technol. **43**:43-50.
40. **Evenblij H.** 2006. Filtration characteristics in membrane bioreactors.
41. **Yang W, Cicek N, Ilg J.** 2006. State-of-the-art of membrane bioreactors: Worldwide research and commercial applications in North America. J. Membr. Sci. **270**:201-211.
42. **Güyer GT, Ince NH.** 2011. Degradation of diclofenac in water by homogeneous and heterogeneous sonolysis. Ultrason. Sonochem. **18**:114-119.
43. **Senta I, Matosic M, Jakopovic HK, Terzic S, Curko J, Mijatovic I, Ahel M.** 2011. Removal of antimicrobials using advanced wastewater treatment. J. Hazard. Mater. **192**:319-328.
44. **Klavarioti M, Mantzavinos D, Kassinos D.** 2009. Removal of residual pharmaceuticals from aqueous systems by advanced oxidation processes. Environ. Int. **35**:402-417.
45. **Fatta-Kassinos D, Vasquez MI, Kümmerer K.** 2011. Transformation products of pharmaceuticals in surface waters and wastewater formed during photolysis and advanced oxidation processes â€" Degradation, elucidation of byproducts and assessment of their biological potency. Chemosphere **85**:693-709.
46. **Müller A, Weiss SC, Beißwenger J, Leukhardt HG, Schulz W, Seitz W, Ruck WKL, Weber WH.** 2012. Identification of ozonation by-products of 4- and 5-methyl-1H-benzotriazole during the treatment of surface water to drinking water. Water Res. **46**:679-690.
47. **Macova M, Escher BI, Reungoat J, Carswell S, Chue KL, Keller J, Mueller JF.** 2010. Monitoring the biological activity of micropollutants during advanced wastewater treatment with ozonation and activated carbon filtration. Water Res. **44**:477-492.
48. **Benner J, Ternes TA.** 2009. Ozonation of Propranolol: Formation of Oxidation Products. Environ. Sci. Technol. **43**:5086-5093.
49. **Hollender J, Zimmermann SG, Koepke S, Krauss M, McArdell CS, Ort C, Singer H, von Gunten U, Siegrist H.** 2009. Elimination of Organic Micropollutants in a Municipal Wastewater Treatment Plant Upgraded with a Full-Scale Post-Ozonation Followed by Sand Filtration. Environ. Sci. Technol. **43**:7862-7869.
50. **Müller E, Schüssler W, Horn H, Lemmer H.** 2013. Aerobic biodegradation of the sulfonamide antibiotic sulfamethoxazole by activated sludge applied as co-substrate and sole carbon and nitrogen source. Chemosphere.
51. **Yang S-F, Lin C-F, Wu C-J, Ng K-K, Yu-Chen Lin A, Andy Hong P-K.** 2012. Fate of sulfonamide antibiotics in contact with activated sludge – Sorption and biodegradation. Water Res. **46**:1301-1308.
52. **Larcher S, Yargeau V.** 2011. Biodegradation of sulfamethoxazole by individual and mixed bacteria. Appl. Microbiol. Biotechnol. **91**:211-218.

References

53. **Loos R, Gawlik BM, Locoro G, Rimaviciute E, Contini S, Bidoglio G.** 2009. EU-wide survey of polar organic persistent pollutants in European river waters. Environ. Pollut. **157**:561-568.
54. **Cano E, Polo JL, La Iglesia A, Bastidas JM.** 2004. A Study on the Adsorption of Benzotriazole on Copper in Hydrochloric Acid Using the Inflection Point of the Isotherm. Adsorption **10**:219-225.
55. **Richardson SD.** 2010. Environmental mass spectrometry: emerging contaminants and current issues. Anal. Chem. **82**:4742-4774.
56. **Weiss S, Reemtsma T.** 2005. Determination of benzotriazole corrosion inhibitors from aqueous environmental samples by liquid chromatography-electrospray ionization-tandem mass spectrometry. Anal. Chem. **77**:7415-7420.
57. **Rubim JC, Gutz IGR, Sala O.** 1983. Surface-enhanced Raman spectra of benzotriazole adsorbed on a silver electrode. J. Mol. Struct. **101**:1-6.
58. **Cotton JB, Scholes IR.** 1967. Benzotriazole and Related Compounds as Corrosion Inhibitors For Copper. British Corrosion Journal **2**:1-5.
59. **Zheludkevich ML, Yasakau KA, Poznyak SK, Ferreira MGS.** 2005. Triazole and thiazole derivatives as corrosion inhibitors for AA2024 aluminium alloy. Corros. Sci. **47**:3368-3383.
60. **Finšgar M, Milošev I.** 2010. Inhibition of copper corrosion by 1,2,3-benzotriazole: A review. Corros. Sci. **52**:2737-2749.
61. **Kim JJ, Kim S-K, Bae J-U.** 2002. Investigation of copper deposition in the presence of benzotriazole. Thin Solid Films **415**:101-107.
62. **Czicholl F.** 2013. Cimachem GmbH, Kirchheimbolanden, Germany, www.chimachem.de.
63. **Janna H, Scrimshaw MD, Williams RJ, Churchley J, Sumpter JP.** 2011. From Dishwasher to Tap? Xenobiotic Substances Benzotriazole and Tolyltriazole in the Environment. Environ. Sci. Technol. **45**:3858-3864.
64. **Cheng C, Phipps D, Alkhaddar RM.** 2005. Treatment of spent metalworking fluids. Water Res. **39**:4051-4063.
65. **Zhang Z, Ren N, Li YF, Kunisue T, Gao D, Kannan K.** 2011. Determination of benzotriazole and benzophenone UV filters in sediment and sewage sludge. Environ. Sci. Technol. **45**:3909-3916.
66. **Farré MI, Pérez S, Kantiani L, Barceló D.** 2008. Fate and toxicity of emerging pollutants, their metabolites and transformation products in the aquatic environment. TrAC Trends in Analytical Chemistry **27**:991-1007.
67. **Breedveld GD, Roseth R, Sparrevik M, Hartnik T, Hem LJ.** 2003. Persistence of the de-icing additive benzotriazole at an abandoned airport. Water, Air, Soil Pollut. Focus **3**:91-101.
68. **Katz J.** 1975. General purpose developer. Google Patents.
69. **Balbo A, Chiavari C, Martini C, Monticelli C.** 2012. Effectiveness of corrosion inhibitor films for the conservation of bronzes and gilded bronzes. Corros. Sci. **59**:204-212.
70. **FLEMING A.** 1944. THE DISCOVERY OF PENICILLIN. Br. Med. Bull. **2**:4-5.
71. **Bennett JW, Chung K-T.** 2001. Alexander Fleming and the discovery of penicillin, p. 163-184, Adv. Appl. Microbiol., vol. Volume 49. Academic Press.
72. **Begley CG, Ellis LM.** 2012. Drug development: Raise standards for preclinical cancer research. Nature **483**:531-533.
73. **Daumerie D, Savioli L, Crompton DDWT, Peters P.** 2010. Working to overcome the global impact of neglected tropical diseases: first WHO report on neglected tropical diseases, vol. 1. World Health Organization.

74. **Hur J, Jawale C, Lee JH.** 2012. Antimicrobial resistance of Salmonella isolated from food animals: A review. Food Res. Int. **45**:819-830.

75. **Montgomery AB, Feigal Jr DW, Sattler F, Mason GR, Catanzaro A, Edison R, Markowitz N, Johnson E, Ogawa S, Rovzar M.** 2013. Pentamidine aerosol versus trimethoprim-sulfamethoxazole for Pneumocystis carinii in acquired immune deficiency syndrome. Am. J. Respir. Crit. Care Med. **151**.

76. **Jia J, Zhu F, Ma X, Cao ZW, Li YX, Chen YZ.** 2009. Mechanisms of drug combinations: interaction and network perspectives. Nature Reviews Drug Discovery **8**:111-128.

77. **Gales AC.** 2012. Burkholderia cepacia complex: evaluation of antimicrobial susceptible profile and characterization of the mechanisms of resistance to sulfamethoxazole/trimethoprim.

78. **Sköld O.** 2010. Sulfonamides and trimethoprim. Expert Rev. Anti Infect. Ther. **8**:1-6.

79. **Zhang X-X, Zhang T, Fang HP.** 2009. Antibiotic resistance genes in water environment. Appl. Microbiol. Biotechnol. **82**:397-414.

80. **Yun M-K, Wu Y, Li Z, Zhao Y, Waddell MB, Ferreira AM, Lee RE, Bashford D, White SW.** 2012. Catalysis and sulfa drug resistance in dihydropteroate synthase. Science **335**:1110-1114.

81. **Jechalke S, Heuer H, Smalla K.** 2013. Antibiotika resistenzgene im Ackerboden. Biospektrum **19**:243-246.

82. **Asimakopoulos AG, Ajibola A, Kannan K, Thomaidis NS.** 2013. Occurrence and removal efficiencies of benzotriazoles and benzothiazoles in a wastewater treatment plant in Greece. Sci. Total Environ. **452–453**:163-171.

83. **Vetter W, Lorenz J.** 2013. Determination of benzotriazoles in dishwasher tabs from Germany and estimation of the discharge into German waters. Environ Sci Pollut R **20**:4435-4440.

84. **Kiss A, Fries E.** 2009. Occurrence of benzotriazoles in the rivers Main, Hengstbach, and Hegbach (Germany). Environ. Sci. Pollut. Res. Int. **16**:702-710.

85. **Reemtsma T, Miehe U, Duennbier U, Jekel M.** 2010. Polar pollutants in municipal wastewater and the water cycle: occurrence and removal of benzotriazoles. Water Res. **44**:596-604.

86. **Verheyen V, Cruickshank A, Wild K, Heaven MW, McGee R, Watkins M, Nash D.** 2009. Soluble, semivolatile phenol and nitrogen compounds in milk-processing wastewaters. J. Dairy Sci. **92**:3484-3493.

87. **Sulej AM, Polkowska Ż, Namieśnik J.** 2011. Pollutants in Airport Runoff Waters. Crit. Rev. Environ. Sci. Technol. **42**:1691-1734.

88. **McNeill KS, Cancilla DA.** 2009. Detection of Triazole Deicing Additives in Soil Samples from Airports with Low, Mid, and Large Volume Aircraft Deicing Activities. Bull. Environ. Contam. Toxicol. **82**:265-269.

89. **Liu Y-S, Ying G-G, Shareef A, Kookana RS.** 2013. Biodegradation of three selected benzotriazoles in aquifer materials under aerobic and anaerobic conditions. J Contam. Hydrol. **151**:131-139.

90. **Liu Y-S, Ying G-G, Shareef A, Kookana RS.** 2011. Biodegradation of three selected benzotriazoles under aerobic and anaerobic conditions. Water Res. **45**:5005-5014.

91. **Radke M, Lauwigi C, Heinkele G, Mürdter TE, Letzel M.** 2009. Fate of the Antibiotic Sulfamethoxazole and Its Two Major Human Metabolites in a Water Sediment Test. Environ. Sci. Technol. **43**:3135-3141.

92. **Hruska K, Franek M.** 2012. Sulfonamides in the environment: a review and a case report. Vet. Med. (Praha) **57**:1-35.

93. **Richardson SD, Ternes TA.** 2005. Water analysis: emerging contaminants and current issues. Anal. Chem. (Wash.) **77**:3807-3838.

94. **Kiss A, Fries E.** 2012. Seasonal source influence on river mass flows of benzotriazoles. J. Environ. Monit. **14**:697-703.
95. **Maeng SK, Sharma SK, Lekkerkerker-Teunissen K, Amy GL.** 2011. Occurrence and fate of bulk organic matter and pharmaceutically active compounds in managed aquifer recharge: A review. Water Res. **45**:3015-3033.
96. **LaPara TM, Burch TR, McNamara PJ, Tan DT, Yan M, Eichmiller JJ.** 2011. Tertiary-Treated Municipal Wastewater is a Significant Point Source of Antibiotic Resistance Genes into Duluth-Superior Harbor. Environ. Sci. Technol. **45**:9543-9549.
97. **Forrest F, Lorenz K, Thompson T, Keenliside J, Kendall J, Charest J.** 2011. A scoping study of livestock antimicrobials in agricultural streams of Alberta. Canadian Water Resources Journal **36**:1-16.
98. **Pedrouzo M, Borrull F, Pocurull E, Marcé RM.** 2011. Presence of pharmaceuticals and hormones in waters from sewage treatment plants. Water Air Soil Poll **217**:267-281.
99. **Reemtsma T, Weiss S, Mueller J, Petrovic M, Gonzalez S, Barcelo D, Ventura F, Knepper TP.** 2006. Polar Pollutants Entry into the Water Cycle by Municipal Wastewater: A European Perspective. Environ. Sci. Technol. **40**:5451-5458.
100. **Yang S-F, Lin C-F, Yu-Chen Lin A, Andy Hong P-K.** 2011. Sorption and biodegradation of sulfonamide antibiotics by activated sludge: Experimental assessment using batch data obtained under aerobic conditions. Water Res. **45**:3389-3397.
101. **Sahar E, Ernst M, Godehardt M, Hein A, Herr J, Kazner C, Melin T, Cikurel H, Aharoni A, Messalem R, Brenner A, Jekel M.** 2011. Comparison of two treatments for the removal of selected organic micropollutants and bulk organic matter: conventional activated sludge followed by ultrafiltration versus membrane bioreactor. Water Sci. Technol. **63**:733-740.
102. **Liu Y-S, Ying G-G, Shareef A, Kookana RS.** 2012. Occurrence and removal of benzotriazoles and ultraviolet filters in a municipal wastewater treatment plant. Environ. Pollut. **165**:225-232.
103. **Larcher S, Yargeau V.** 2012. Biodegradation of sulfamethoxazole: current knowledge and perspectives. Appl. Microbiol. Biotechnol. **96**:309-318.
104. **Bouju H, Ricken B, Beffa T, Corvini PF, Kolvenbach BA.** 2012. Isolation of bacterial strains capable of sulfamethoxazole mineralization from an acclimated membrane bioreactor. Appl. Environ. Microbiol. **78**:277-279.
105. **Gauthier H, Yargeau V, Cooper DG.** 2010. Biodegradation of pharmaceuticals by Rhodococcus rhodochrous and Aspergillus niger by co-metabolism. Sci. Total Environ. **408**:1701-1706.
106. **Eibes G, Debernardi G, Feijoo G, Moreira MT, Lema J.** 2011. Oxidation of pharmaceutically active compounds by a ligninolytic fungal peroxidase. Biodegradation **22**:539-550.
107. **Cavaluzzi MJ, Borer PN.** 2004. Revised UV extinction coefficients for nucleoside5'monophosphates and unpaired DNA and RNA. Nucleic Acids Res. **32**:e13-e13.
108. **Jover E, Matamoros V, Bayona JM.** 2009. Characterization of benzothiazoles, benzotriazoles and benzosulfonamides in aqueous matrixes by solid-phase extraction followed by comprehensive two-dimensional gas chromatography coupled to time-of-flight mass spectrometry. J. Chromatogr. **1216**:4013-4019.
109. **Nakata H, Shinohara R-I, Nakazawa Y, Isobe T, Sudaryanto A, Subramanian A, Tanabe S, Zakaria MP, Zheng GJ, Lam PKS, Kim EY, Min B-Y, We S-U, Viet PH, Tana TS, Prudente M, Frank D, Lauenstein G, Kannan K.** 2012. Asia–Pacific mussel watch for emerging pollutants: Distribution of synthetic musks and benzotriazole UV stabilizers in Asian and US coastal waters. Mar. Pollut. Bull. **64**:2211-2218.

110. Wolschke H, Xie Z, Möller A, Sturm R, Ebinghaus R. 2011. Occurrence, distribution and fluxes of benzotriazoles along the German large river basins into the North Sea. Water Res. **45**:6259-6266.

111. Cancilla DA, Martinez J, Van Aggelen GC. 1998. Detection of aircraft deicing/antiicing fluid additives in a perched water monitoring well at an international airport. Environ. Sci. Technol. **32**:3834-3835.

112. Seeland A, Oetken M, Kiss A, Fries E, Oehlmann J. 2012. Acute and chronic toxicity of benzotriazoles to aquatic organisms. Environ Sci Pollut R **19**:1781-1790.

113. **Pedrazzani R, Ceretti E, Zerbini I, Casale R, Gozio E, Bertanza G, Gelatti U, Donato F, Feretti D.** 2012. Biodegradability, toxicity and mutagenicity of detergents: Integrated experimental evaluations. Ecotoxicol. Environ. Saf. **84**:274-281.

114. Cancilla DA, Holtkamp A, Matassa L, Fang X. 1997. Isolation and characterization of Microtox®-active components from aircraft de-icing/anti-icing fluids. Environ. Toxicol. Chem. **16**:430-434.

115. Chapra SC. 1997. Surface water-quality modeling, vol. 1. McGraw-Hill New York.

116. **Schultz-Fademrecht C, Wichern M, Horn H.** 2008. The impact of sunlight on inactivation of indicator microorganisms both in river water and benthic biofilms. Water Res. **42**:4771-4779.

117. **Corsi SR, Geis SW, Loyo-Rosales JE, Rice CP, Sheesley RJ, Failey GG, Cancilla DA.** 2006. Characterization of aircraft deicer and anti-icer components and toxicity in airport snowbanks and snowmelt runoff. Environ. Sci. Technol. **40**:3195-3202.

118. **Voutsa D, Hartmann P, Schaffner C, Giger W.** 2006. Benzotriazoles, alkylphenols and bisphenol A in municipal wastewaters and in the Glatt River, Switzerland. Environ Sci Pollut R **13**:333-341.

119. Gruden CL, Dow SM, Hernandez MT. 2001. Fate and toxicity of aircraft deicing fluid additives through anaerobic digestion. Water Environ. Res. **73**:72-79.

120. **Zhang Z, Ren N, Li Y-F, Kunisue T, Gao D, Kannan K.** 2011. Determination of Benzotriazole and Benzophenone UV Filters in Sediment and Sewage Sludge. Environ. Sci. Technol. **45**:3909-3916.

121. Hart DS, Davis LC, Erickson LE, Callender TM. 2004. Sorption and partitioning parameters of benzotriazole compounds. Microchem. J. **77**:9-17.

122. Liu Y-S, Ying G-G, Shareef A, Kookana RS. 2011. Biodegradation of three selected benzotriazoles under aerobic and anaerobic conditions. Water Res. **45**:5005-5014.

123. Nakata H, Murata S, Filatreau J. 2009. Occurrence and concentrations of benzotriazole UV stabilizers in marine organisms and sediments from the Ariake Sea, Japan. Environ. Sci. Technol. **43**:6920-6926.

124. Kadar E, Dashfield S, Hutchinson TH. 2010. Developmental toxicity of benzotriazole in the protochordate Ciona intestinalis (Chordata, Ascidiae). Anal. Bioanal. Chem. **396**:641-647.

125. Cancilla DA, Holtkamp A, Matassa L, Fang XC. 1997. Isolation and characterization of microtox(R)-active components from aircraft de-icing/anti-icing fluids. Environ. Toxicol. Chem. **16**:430-434.

126. **Durjava MK, Kolar B, Arnus L, Papa E, Kovarich S, Sahlin U, Peijnenburg W.** 2013. Experimental assessment of the environmental fate and effects of triazoles and benzotriazole. Alternatives to laboratory animals : ATLA **41**:65-75.

127. **Weiss S, Jakobs J, Reemtsma T.** 2006. Discharge of three benzotriazole corrosion inhibitors with municipal wastewater and improvements by membrane bioreactor treatment and ozonation. Environ. Sci. Technol. **40**:7193-7199.

128. **Baumann U, Schefer W.** 1990. Einfaches Testsystem zur Beurteilung der anaeroben Abbaubarkeit organischer Stoffe. Textilveredlung **25**:248-251.

129. **Cervantes FJ, Gutiérrez CH, López KY, Estrada-Alvarado MI, Meza-Escalante ER, Texier A-C, Cuervo F, Gómez J.** 2008. Contribution of quinone-reducing microorganisms to the anaerobic biodegradation of organic compounds under different redox conditions. Biodegradation **19**:235-246.

130. **Lesjean B, Gnirss R, Buisson H, Keller S, Tazi-Pain A, Luck F.** 2005. Outcomes of a 2-year investigation on enhanced biological nutrients removal and trace organics elimination in membrane bioreactor (MBR). Water Sci. Technol. **52**:453-460.

131. **Radjenovic J, Petrovic M, Barcelo D.** 2009. Fate and distribution of pharmaceuticals in wastewater and sewage sludge of the conventional activated sludge (CAS) and advanced membrane bioreactor (MBR) treatment. Water Res. **43**:831-841.

132. **Buitrón G, Capdeville B.** 1995. Enhancement of the biodegradation activity by the acclimation of the inoculum. Environ. Technol. **16**:1175-1184.

133. **Petrovic M, de Alda MJL, Diaz-Cruz S, Postigo C, Radjenovic J, Gros M, Barcelo D.** 2009. Fate and removal of pharmaceuticals and illicit drugs in conventional and membrane bioreactor wastewater treatment plants and by riverbank filtration. Philos T R SOC A **367**:3979-4003.

134. **Drillia P, Dokianakis SN, Fountoulakis MS, Kornaros M, Stamatelatou K, Lyberatos G.** 2005. On the occasional biodegradation of pharmaceuticals in the activated sludge process: The example of the antibiotic sulfamethoxazole. J. Hazard. Mater. **122**:259-265.

135. **Matamoros V, Jover E, Bayona JM.** 2010. Occurrence and fate of benzothiazoles and benzotriazoles in constructed wetlands. Water science and technology : a journal of the International Association on Water Pollution Research **61**:191-198.

136. **Field JA, Cervantes FJ, van der Zee FP, Lettinga G.** 2000. Role of quinones in the biodegradation of priority pollutants: a review. Water Sci. Technol. **42**:215-222.

137. **Stuart M, Lapworth D, Crane E, Hart A.** 2012. Review of risk from potential emerging contaminants in UK groundwater. Sci. Total Environ. **416**:1-21.

138. **Kadar E, Dashfield S, Hutchinson T.** 2010. Developmental toxicity of benzotriazole in the protochordate Ciona intestinalis (Chordata, Ascidiae). Anal. Bioanal. Chem. **396**:641-647.

139. **Reasoner DJ, Geldreich EE.** 1985. A new medium for the enumeration and subculture of bacteria from potable water. Appl. Environ. Microbiol. **49**:1-7.

140. **Aktaş Ö.** 2012. Effect of S0/X0 ratio and acclimation on respirometry of activated sludge in the cometabolic biodegradation of phenolic compounds. Bioresour. Technol. **111**:98-104.

141. **Çeçen F, Kocamemi B, Aktaş Ö.** 2010. Metabolic and Co-metabolic Degradation of Industrially Important Chlorinated Organics Under Aerobic Conditions, p. 161-178. *In* Fatta-Kassinos D, Bester K, Kümmerer K (ed.), Xenobiotics in the Urban Water Cycle, vol. 16. Springer Netherlands.

142. **Moreno G, Buitrón G.** 2004. Influence of the origin of the inoculum and the acclimation strategy on the degradation of 4-chlorophenol. Bioresour. Technol. **94**:215-218.

143. **Chong N-M, Luong M, Hwu C-S.** 2012. Biogenic substrate benefits activated sludge in acclimation to a xenobiotic. Bioresour. Technol. **104**:181-186.

144. **Tripp HJ, Kitner JB, Schwalbach MS, Dacey JW, Wilhelm LJ, Giovannoni SJ.** 2008. SAR11 marine bacteria require exogenous reduced sulphur for growth. Nature **452**:741-744.

145. **Kaeberlein T, Lewis K, Epstein SS.** 2002. Isolating "uncultivable" microorganisms in pure culture in a simulated natural environment. Science **296**:1127-1129.

146. **Nichols D, Lewis K, Orjala J, Mo S, Ortenberg R, O'Connor P, Zhao C, Vouros P, Kaeberlein T, Epstein SS.** 2008. Short peptide induces an "uncultivable" microorganism to grow in vitro. Appl. Environ. Microbiol. **74**:4889-4897.

147. Bollmann A, Lewis K, Epstein SS. 2007. Incubation of environmental samples in a diffusion chamber increases the diversity of recovered isolates. Appl. Environ. Microbiol. **73**:6386-6390.

148. **Virdis B, Rabaey K, Rozendal RA, Yuan Z, Keller J.** 2010. Simultaneous nitrification, denitrification and carbon removal in microbial fuel cells. Water Res. **44**:2970-2980.

149. **Seo J-S, Keum Y-S, Li QX.** 2009. Bacterial degradation of aromatic compounds. Int. J. Env. Res. Public Health **6**:278-309.

150. **Bromilow RH, Evans AA, Nicholls PH.** 1999. Factors affecting degradation rates of five triazole fungicides in two soil types: 1. Laboratory incubations. Pestic. Sci. **55**:1129-1134.

151. **Patil SG, Nicholls PH, Chamberlain K, Briggs GG, Bromilow RH.** 1988. Degradation rates in soil of 1-benzyltriazoies and two triazole fungicides. Pestic. Sci. **22**:333-342.

152. **Muyzer G, de Waal EC, Uitterlinden AG.** 1993. Profiling of complex microbial populations by denaturing gradient gel electrophoresis analysis of polymerase chain reaction-amplified genes coding for 16S rRNA. Appl. Environ. Microbiol. **59**:695-700.

153. **Parkhomchuk D, Borodina T, Amstislavskiy V, Banaru M, Hallen L, Krobitsch S, Lehrach H, Soldatov A.** 2009. Transcriptome analysis by strand-specific sequencing of complementary DNA. Nucleic Acids Res. **37**:e123.

154. **Peng Y, Leung HCM, Yiu SM, Chin FYL.** 2012. IDBA-UD: a de novo assembler for single-cell and metagenomic sequencing data with highly uneven depth. Bioinformatics **28**:1420-1428.

155. **Trimble W, Keegan K, D'Souza M, Wilke A, Wilkening J, Gilbert J, Meyer F.** 2012. Short-read reading-frame predictors are not created equal: sequence error causes loss of signal. BMC Bioinformatics **13**:183.

156. **Camacho C, Coulouris G, Avagyan V, Ma N, Papadopoulos J, Bealer K, Madden T.** 2009. BLAST+: architecture and applications. BMC Bioinformatics **10**:421.

157. **Langmead B, Salzberg SL.** 2012. Fast gapped-read alignment with Bowtie 2. Nat. Methods **9**:357-359.

158. **Tezel U, Tandukar M, Martinez RJ, Sobecky PA, Pavlostathis SG.** 2012. Aerobic Biotransformation of n-Tetradecylbenzyldimethylammonium Chloride by an Enriched Pseudomonas spp. Community. Environ. Sci. Technol. **46**:8714-8722.

159. **Kalyani DC, Telke AA, Dhanve RS, Jadhav JP.** 2009. Ecofriendly biodegradation and detoxification of Reactive Red 2 textile dye by newly isolated Pseudomonas sp. SUK1. J. Hazard. Mater. **163**:735-742.

160. **Jadhav J, Phugare S, Dhanve R, Jadhav S.** 2010. Rapid biodegradation and decolorization of Direct Orange 39 (Orange TGLL) by an isolated bacterium Pseudomonas aeruginosa strain BCH. Biodegradation **21**:453-463.

161. **Gan HM, Shahir S, Ibrahim Z, Yahya A.** 2011. Biodegradation of 4-aminobenzenesulfonate by Ralstonia sp. PBA and Hydrogenophaga sp. PBC isolated from textile wastewater treatment plant. Chemosphere **82**:507-513.

162. **Sun B, Ko K, Ramsay J.** 2011. Biodegradation of 1,4-dioxane by a Flavobacterium. Biodegradation **22**:651-659.

163. **Janse I, Bok J, Zwart G.** 2004. A simple remedy against artifactual double bands in denaturing gradient gel electrophoresis. J. Microbiol. Methods **57**:279-281.

164. **Sørensen SR, Holtze MS, Simonsen A, Aamand J.** 2007. Degradation and Mineralization of Nanomolar Concentrations of the Herbicide Dichlobenil and Its Persistent Metabolite 2,6-Dichlorobenzamide by Aminobacter spp. Isolated from Dichlobenil-Treated Soils. Appl. Environ. Microbiol. **73**:399-406.

165. **Thompson JR, Marcelino LA, Polz MF.** 2002. Heteroduplexes in mixed-template amplifications: formation, consequence and elimination by 'reconditioning PCR'. Nucleic Acids Res. **30:**2083-2088.

166. **Nocker A, Camper AK.** 2006. Selective removal of DNA from dead cells of mixed bacterial communities by use of ethidium monoazide. Appl. Environ. Microbiol. **72:**1997-2004.

167. **Heuer H, Krsek M, Baker P, Smalla K, Wellington EM.** 1997. Analysis of actinomycete communities by specific amplification of genes encoding 16S rRNA and gel-electrophoretic separation in denaturing gradients. Appl. Environ. Microbiol. **63:**3233-3241.

168. **Muyzer G, Smalla K.** 1998. Application of denaturing gradient gel electrophoresis (DGGE) and temperature gradient gel electrophoresis (TGGE) in microbial ecology. Antonie Van Leeuwenhoek **73:**127-141.

169. **Venter JC, Remington K, Heidelberg JF, Halpern AL, Rusch D, Eisen JA, Wu D, Paulsen I, Nelson KE, Nelson W, Fouts DE, Levy S, Knap AH, Lomas MW, Nealson K, White O, Peterson J, Hoffman J, Parsons R, Baden-Tillson H, Pfannkoch C, Rogers Y-H, Smith HO.** 2004. Environmental Genome Shotgun Sequencing of the Sargasso Sea. Science **304:**66-74.

170. **Liu Y, Zhang T, Fang HHP.** 2005. Microbial community analysis and performance of a phosphate-removing activated sludge. Bioresour. Technol. **96:**1205-1214.

171. **Kümmerer Ke.** 2004. Pharmaceuticals in the environment: Sources, fate, effects, and risks, 2nd edition ed. Springer, Berlin, Heidelberg, Germany.

172. **Kümmerer Ke.** 2008. Pharmaceuticals in the environment, 3rd, revised and enlarged edition ed. Springer.

173. **Baran W, Sochacka J, Wardas W.** 2006. Toxicity and biodegradability of sulfonamides and products of their photocatalytic degradation in aqueous solutions. Chemosphere **65:**1295–1299.

174. **Xu B, Mao D, Luo Y, Xu L.** 2011. Sulfamethoxazole biodegradation and biotransformation in the water–sediment system of a natural river. Bioresour. Technol. **102:**7069-7076.

175. **Heberer T.** 2002. Occurrence, fate, and removal of pharmaceutical residues in the aquatic environment: a review of recent research data. Toxicol. Lett. **131:**5-17.

176. **Ternes T, Joss A.** 2007. Human Pharmaceuticals, Hormones and Fragrances The Challenge of Micropollutants in Urban Water Management.

177. **Kümmerer K.** 2009. Antibiotics in the aquatic environment--a review--part II. Chemosphere **75:**435-441.

178. **Pérez S, Eichhorn P, Aga DS.** 2005. Evaluating the biodegradability of sulfamethazine, sulfamethoxazole, sulfathiazole, and trimethoprim at different stages of sewage treatment. Environ. Toxicol. Chem. **24:**1361-1367.

179. **Agerso Y, Petersen A.** 2007. The tetracycline resistance determinant Tet 39 and the sulphonamide resistance gene sulII are common among resistant Acinetobacter spp. isolated from integrated fish farms in Thailand. J. Antimicrob. Chemother. **59:**23-27.

180. **Szczepanowski R, Linke B, Krahn I, Gartemann K-H, Gützkow T, Eichler W, Pühler A, Schlüter A.** 2009. Detection of 140 clinically relevant antibiotic-resistance genes in the plasmid metagenome of wastewater treatment plant bacteria showing reduced susceptibility to selected antibiotics. Microbiology **155:**2306-2319.

181. **Cavallucci S.** 2007. Top 200: What's topping the charts in prescription drugs this year. Pharmacy Practice, Canadian Healthcare Network. Available from http://www.imshealthcanada.com/vgn/images/portal/cit_40000873/13/31/8286270612-TOP200-07-final.pdf.

182. **Benotti MJ, Trenholm RA, Vanderford BJ, Holady JC, Stanford BD, Snyder SA.** 2008. Pharmaceuticals and endocrine disrupting compounds in US drinking water. Environ. Sci. Technol. **43**:597-603.

183. **Miège C, Choubert J, Ribeiro L, Eusèbe M, Coquery M.** 2009. Fate of pharmaceuticals and personal care products in wastewater treatment plants-Conception of a database and first results. Environ. Pollut. **157**:1721-1726.

184. **Sacher F, Lange FT, Brauch HJ, Blankenhorn I.** 2001. Pharmaceuticals in groundwaters: Analytical methods and results of a monitoring program in Baden-Wurttemberg, Germany. J. Chromatogr. **938**:199-210.

185. **Onesios K, Yu J, Bouwer E.** 2009. Biodegradation and removal of pharmaceuticals and personal care products in treatment systems: a review. Biodegradation **20**:441-466.

186. **Huang T-S, Kunin CM, Yan B-S, Chen Y-S, Lee SS-J, Syu W.** 2012. Susceptibility of Mycobacterium tuberculosis to sulfamethoxazole, trimethoprim and their combination over a 12 year period in Taiwan. J. Antimicrob. Chemother. **67**:633-637.

187. **Fajardo A, Martínez JL.** 2008. Antibiotics as signals that trigger specific bacterial responses. Curr. Opin. Microbiol. **11**:161-167.

188. **Jiang X, Shi L.** 2013. Distribution of tetracycline and trimethoprim/sulfamethoxazole resistance genes in aerobic bacteria isolated from cooked meat products in Guangzhou, China. Food Control **30**:30-34.

189. **Liu F, Wu J, Ying G-G, Luo Z, Feng H.** 2012. Changes in functional diversity of soil microbial community with addition of antibiotics sulfamethoxazole and chlortetracycline. Appl. Microbiol. Biotechnol. **95**:1615-1623.

190. **Gutiérrez I, Watanabe N, Harter T, Glaser B, Radke M.** 2010. Effect of sulfonamide antibiotics on microbial diversity and activity in a Californian Mollic Haploxeralf. J. Soils Sed. **10**:537-544.

191. **Collado N, Buttiglieri G, Marti E, Ferrando-Climent L, Rodriguez-Mozaz S, Barceló D, Comas J, Rodriguez-Roda I.** Effects on activated sludge bacterial community exposed to sulfamethoxazole. Chemosphere.

192. **Göbel A, McArdell CS, Joss A, Siegrist H, Giger W.** 2007. Fate of sulfonamides, macrolides, and trimethoprim in different wastewater treatment technologies. Sci. Total Environ. **372**:361-371.

193. **Niu J, Zhang L, Li Y, Zhao J, Lv S, Xiao K.** 2013. Effects of environmental factors on sulfamethoxazole photodegradation under simulated sunlight irradiation: Kinetics and mechanism. Journal of Environmental Sciences **25**:1098-1106.

194. **Trovó AG, Nogueira RFP, Agüera A, Sirtori C, Fernández-Alba AR.** 2009. Photodegradation of sulfamethoxazole in various aqueous media: Persistence, toxicity and photoproducts assessment. Chemosphere **77**:1292-1298.

195. **Hyland KC, Dickenson ERV, Drewes JE, Higgins CP.** 2012. Sorption of ionized and neutral emerging trace organic compounds onto activated sludge from different wastewater treatment configurations. Water Res. **46**:1958-1968.

196. **Joss A, Zabczynski S, Gobel A, Hoffmann B, Loffler D, McArdell CS, Ternes TA, Thomsen A, Siegrist H.** 2006. Biological degradation of pharmaceuticals in municipal wastewater treatment: Proposing a classification scheme. Water Res. **40**:1686-1696.

197. **Alexy R, Kümpel T, Kümmerer K.** 2004. Assessment of degradation of 18 antibiotics in the Closed Bottle Test. Chemosphere **57**:505-512.

198. **Weisburg WG, Barns SM, Pelletier DA, Lane DJ.** 1991. 16S ribosomal DNA amplification for phylogenetic study. J. Bacteriol. **173**:697-703.

199. **Ludwig W, Strunk O, Westram R, Richter L, Meier H, Yadhukumar, Buchner A, Lai T, Steppi S, Jobb G, Förster W, Brettske I, Gerber S, Ginhart AW, Gross O, Grumann S, Hermann S, Jost R, König A, Liss T, Lüßmann R, May M, Nonhoff B, Reichel B, Strehlow R, Stamatakis A, Stuckmann N, Vilbig A, Lenke M, Ludwig**

T, Bode A, Schleifer KH. 2004. ARB: a software environment for sequence data. Nucleic Acids Res. **32:**1363-1371.

200. **Ludwig W, Klenk H-P.** 2001. Overview: A Phylogenetic Backbone and Taxonomic Framework for Procaryotic Systematics, p. 49-65. *In* Boone D, Castenholz R (ed.), Bergey's Manual® of Systematic Bacteriology. Springer New York.

201. **Jiang Q, Fu B, Chen Y, Wang Y, Liu H.** 2013. Quantification of viable but nonculturable bacterial pathogens in anaerobic digested sludge. Appl. Microbiol. Biotechnol. **97:**6043-6050.

202. **Wagner M, Aßmus B, Hartmann A, Hutzler P, Amann R.** 1994. In situ analysis of microbial consortia in activated sludge using fluorescently labelled, rRNA-targeted oligonucleotide probes and confocal scanning laser microscopy. J. Microsc. **176:**181-187.

203. **Vartoukian SR, Palmer RM, Wade WG.** 2010. Strategies for culture of 'unculturable' bacteria. FEMS Microbiol. Lett. **309:**1-7.

204. **Wagner M, Amann R, Lemmer H, Schleifer KH.** 1993. Probing activated sludge with oligonucleotides specific for proteobacteria: inadequacy of culture-dependent methods for describing microbial community structure. Appl. Environ. Microbiol. **59:**1520-1525.

205. **Snaidr J, Amann R, Huber I, Ludwig W, Schleifer KH.** 1997. Phylogenetic analysis and in situ identification of bacteria in activated sludge. Appl. Environ. Microbiol. **63:**2884-2896.

206. **Chiellini C, Munz G, Petroni G, Lubello C, Mori G, Verni F, Vannini C.** 2013. Characterization and Comparison of Bacterial Communities Selected in Conventional Activated Sludge and Membrane Bioreactor Pilot Plants: A Focus on Nitrospira and Planctomycetes Bacterial Phyla. Curr. Microbiol. **67:**77-90.

207. **Wells GF, Park H-D, Eggleston B, Francis CA, Criddle CS.** 2011. Fine-scale bacterial community dynamics and the taxa–time relationship within a full-scale activated sludge bioreactor. Water Res. **45:**5476-5488.

208. **De Gusseme B, Vanhaecke L, Verstraete W, Boon N.** 2011. Degradation of acetaminophen by Delftia tsuruhatensis and Pseudomonas aeruginosa in a membrane bioreactor. Water Res. **45:**1829-1837.

209. **Shiomi N, Ako M.** 2012. Biodegradation of Melamine and Cyanuric Acid by a Newly-Isolated Microbacterium Strain. Advances in Microbiology **2:**303-309.

210. **Chunming W, Chunlian LIDW.** 2009. Biodegradation of Naphthalene, Phenanthrene, Anthracene and Pyrene by Microbacterium sp. 3-28. Chin. J. Appl. Environ. Biol. **3:**017.

211. **Satola B, Wübbeler J, Steinbüchel A.** 2013. Metabolic characteristics of the species Variovorax paradoxus. Appl. Microbiol. Biotechnol. **97:**541-560.

212. **Islas-Espinoza M, Reid B, Wexler M, Bond P.** 2012. Soil Bacterial Consortia and Previous Exposure Enhance the Biodegradation of Sulfonamides from Pig Manure. Microb. Ecol. **64:**140-151.

213. **Cohen GN.** 2011. Bacterial Growth, p. 1-10, Microbial Biochemistry. Springer Netherlands.

i want morebooks!

Buy your books fast and straightforward online - at one of world's fastest growing online book stores! Environmentally sound due to Print-on-Demand technologies.

Buy your books online at
www.get-morebooks.com

Kaufen Sie Ihre Bücher schnell und unkompliziert online – auf einer der am schnellsten wachsenden Buchhandelsplattformen weltweit! Dank Print-On-Demand umwelt- und ressourcenschonend produziert.

Bücher schneller online kaufen
www.morebooks.de

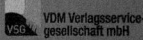 VDM Verlagsservicegesellschaft mbH
Heinrich-Böcking-Str. 6-8   Telefon: +49 681 3720 174   info@vdm-vsg.de
D - 66121 Saarbrücken   Telefax: +49 681 3720 1749   www.vdm-vsg.de

Printed by Books on Demand GmbH, Norderstedt / Germany